Opening the Mind's Eye

ALSO BY IAN ROBERTSON

Mind Sculpture: Unlocking Your Brain's Untapped Potential

Opening the Mind's Eye

How Images and Language Teach Us How to See

IAN ROBERTSON

St. Martin's Press New York

 For information, address St. Martin's Press, 175 Fifth Avenue, New York, N.Y. 10010.

www.stmartins.com

Library of Congress Cataloging-in-Publication Data

Robertson, Ian H.
Opening the mind's eye : how images and language teach us how to see / Ian Robertson.— 1st U.S. ed.
p. cm.
Originally published: The mind's eye. London : Bantam, 2002.
ISBN 0-312-30657-1
1. Imagery (Psychology) 2. Visualization. I. Title.
BF367 .R63 2003
153.3'2—dc21

2002151155

First published in Great Britain in 2002 under the title *The Mind's Eye* by Bantam Press.

First U.S. Edition: March 2003

10 9 8 7 6 5 4 3 2 1

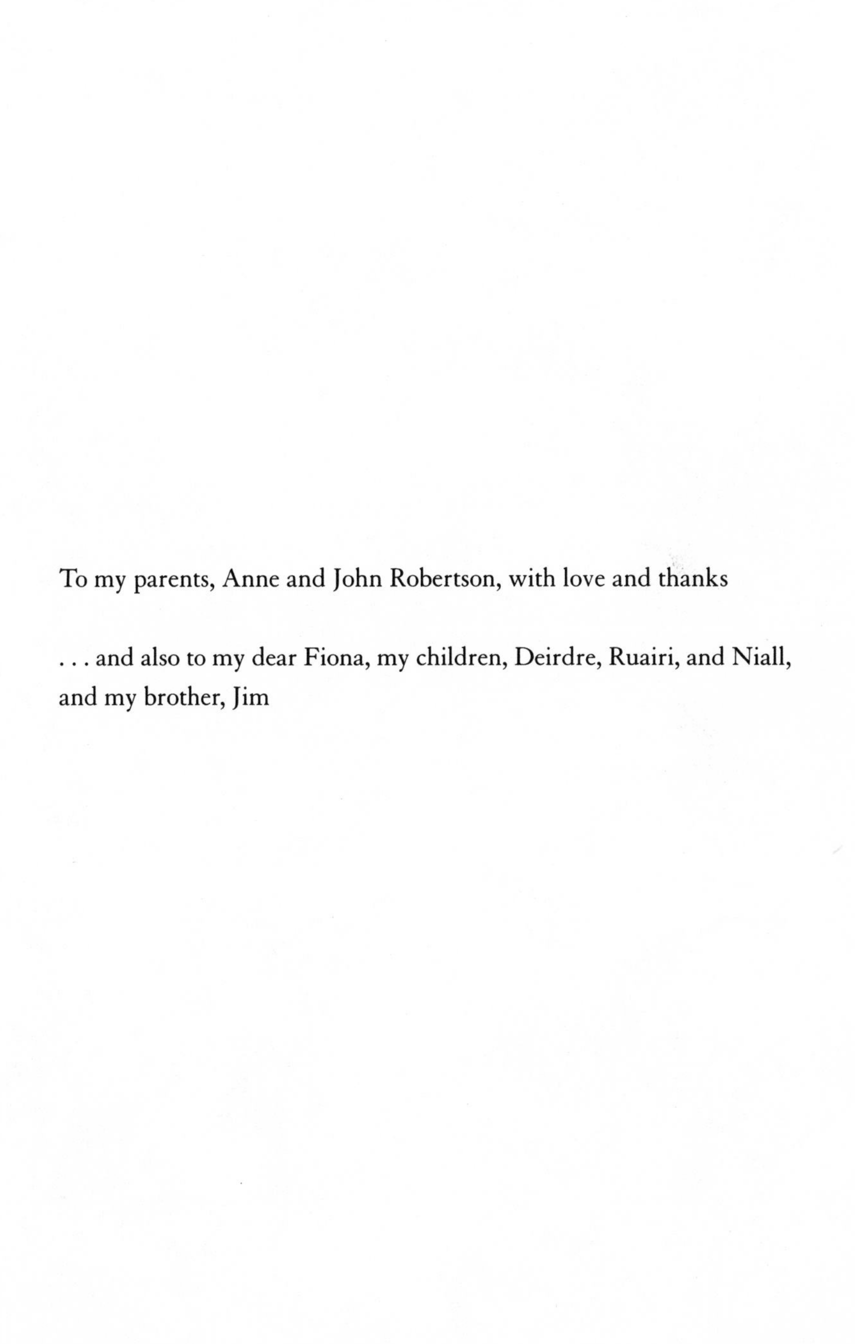

To my parents, Anne and John Robertson, with love and thanks

. . . and also to my dear Fiona, my children, Deirdre, Ruairi, and Niall, and my brother, Jim

Contents

Acknowledgments

My sincere thanks to Sally Gaminara and Simon Thorogood of Transworld Publishers for their help and enthusiasm while writing this book, and also to my agent Felicity Bryan and her assistant Michele Topham. Without the coal fires, scalding coffee, and consoling rock buns of Bewley's Café in Westmoreland Street in Dublin, I would not have survived the early morning prework writing sessions there that produced this book.

The Vividness of Visual Imagery questionnaire in Chapter 4 is reproduced with the kind permission of Professor David Marks and with permission from the *British Journal of Psychology,* copyright the British Psychological Society. Quotations in the introduction to Chapter 4 are reprinted from *Classic Cases in Neuropsychology,* by permission of Psychology Press Limited, Hove, UK. "The Cool Web" by Robert Graves is reproduced with the permission of Carcanet Press Ltd: Robert Graves's *Complete Poems* (1995). The drawing of the horse on page 9 is reproduced from *Nadia: A Case of Extraordinary Drawing Ability in an Autistic*

Child by Lorna Selfe (1977), page 17, by permission of the publisher Academie Press. Reproduction of the "croquet" figure on page 76 is by permission of Psychonomic Society Publications, *Memory and Cognition*, Vol. 16, part 3.

Opening the Mind's Eye

I

A Word in Your Eye

The Cool Web

Children are dumb to say how hot the day is,
How hot the scent is of the summer rose,
How dreadful the black wastes of evening sky,
How dreadful the tall soldiers drumming by.

But we have speech, to chill the angry day,
And speech, to dull the rose's cruel scent.
We spell away the overhanging night,
We spell away the soldiers and the fright.

There's a cool web of language winds us in,
Retreat from too much joy or too much fear:
We grow sea-green at last and coldly die
In brininess and volubility.

But if we let our tongues lose self-possession,
Throwing off language and its watery clasp
Before our death, instead of when death comes,
Facing the wide glare of the children's day,
Facing the rose, the dark sky and the drums,
We shall go mad no doubt and die that way.

—Robert Graves

Western societies have largely lost the ability to think in images rather than words. That, in a nutshell, is the argument of this book. In his poem "The Cool Web," Robert Graves makes the point very elegantly, and as you'll see if you read on, modern neuroscience backs him up.

Take a moment to think about the last time you ate an apple. When was it? Where were you? What kind of apple was it? It is likely that, as you did this, you relied on both words and images. But for many of you the images would have been pretty bloodless, and you probably re-created that event to a great extent with words—"Oh, I think it was on Sunday, and I was in the kitchen after lunch . . . it was a red apple."

Now try to recall this event in a quite different way. Close your eyes and try to see the apple in your mind's eye. Try to visualize its color, the blemishes on its skin—the tilt of the stalk. Now imagine feeling the apple—its texture, little indentations, the odd bruise, the sheer hard, smooth roundness of it. Try to taste it next. Imagine its waxy, brittle skin yielding to your teeth, the sweet, acidy juices flowing over your tongue, the dissolving of the flesh into soft flakes, and the sensation of swallowing. Finally, hear the apple—the juicy crunch as you break it with your teeth, the sound of your own chewing inside your head.

Visualizing eating an apple in this way is very different from remembering it casually as an event. It's as different as someone telling you about the taste of some exotic tropical fruit compared with tasting it yourself. Yet it is the nature of words that they tend to transform experiences into a rather bloodless code that can starve our brains of the rich images that wordless imagining can evoke.

It's artificial, of course, to separate words and images like this. Poems like "The Cool Web" work precisely because the words trigger images as well as other word-thoughts. Yet most of us, most of the time—at work, home, watching TV, reading newspapers, studying, sitting in a traffic jam—don't think in images nearly enough. Why should we? Language is the great achievement of evolution—an essential ingredient in what makes human beings unique on the planet. But there are costs to the way we have grown dependent on the spoken and written word.

Imagery consists of the mental sights, sounds, smells, tastes, touch, and other bodily sensations that we can re-create with incredible vividness in that private, infinite universe within our skulls. The human brain is the most complex object in the known universe and it has the most incredible abilities, some of which—like imagery—are underused.

Imagery is important, but in Western culture, language is king. In school we steadily wrap our children's brains in the "cool web of language"—it would be terrible if we didn't, but there is a cost to everything. By neglecting imagery we risk the withering of a whole set of quite remarkable mental capacities. In this book I will give you the scientific evidence to back up these arguments, but I will also give you many exercises in imagery to try out. These exercises are designed to illustrate how the mind's eye works and to help you assess how well you can use it and what effects using it can have on your mind and body.

Children think mostly in images before word-dominated school clouds their mind's eye. That's why this book begins where Robert Graves's poem begins—with the child's mind and its sometimes joyful, sometimes terrifying, image-filled world, untamed by words. Why do most of us lose this powerful way of thinking as we grow up? And why is it we remember so little from before the age of four?

One consequence of the clouding over of the mind's eye is that we only "see" a fraction of what is before our eyes. Most of the time we see, hear, feel, taste, and smell what our brains *expect* rather than the sensations themselves. Much modern art tries to shock or surprise us out of these image-clouding mental habits into *seeing* more purely with the mind's eye, uncluttered by well-worn categories and labels. When we cultivate imagery and visualization in the mind's eye, we use parts of our brain that are not triggered by verbal thoughts. But the moment we speak or think in words, we sabotage this power of the mind's eye. I'll show you in Chapter 6, for example, how self-professed but amateur wine connoisseurs can't tell wines apart if they talk about the wine while drinking it, but they can if they stay silent and let the taste imagery linger in their mind, unfettered by words.

Neuroscientists can now watch the mind's eye at work in the brain and see how it uses quite different parts of the brain from those we use for other types of thinking and remembering. This research reveals that the right half of our brain—which has a limited way with words—can "know" things but be unable to "say" them. You can, for instance, be good at visualizing the color scheme of your new house but bad at working out in your mind whether the sofa will fit in the alcove: different parts of the brain control these different workings of the mind's eye.

In Chapter 3—"How Your Brain Creates Images"—I get down to the business of helping you assess how well you can visualize. Are you a verbalizer or visualizer; do you think mainly in words or images? How well can you mentally picture your best friend's face? Or the details of your front door? Can you imagine vividly the sound of a violin playing? How clearly can you "feel" the imagined touch of someone's finger stroking your cheek? We all vary in how vividly we can create mental pictures in all the senses.

The more vivid a visualizer you are, then, on average, the better you will be at—to give a few examples—remembering your dreams, succumbing to hypnosis, and thinking creatively. The good news is that even if you are a poor visualizer you can train yourself to be better: deaf people, for instance, who learn a sign language that forces them to use mental maps and the mind's eye become much better at thinking in images. London taxi drivers have to learn the spatial layout of London perfectly so that they can create the shortest route from any point of London to any other. A key brain area—the hippocampus—is enlarged in taxi drivers who have used their mind's eye in this way for many years, compared with their younger, less experienced colleagues. In other words, you can train your own mind by practicing imagery, and the great thing about visualization is that you can do it anywhere—from the dentist's waiting room to sitting in a traffic jam.

The better you can use your mind's eye, the more creative you are likely to be: in Chapter 5—"Better Imagery—More Creativity"—you'll see that Albert Einstein went to a school that taught children to think in visual images. At the age of sixteen he used visual imagery to carry out a breakthrough "thought experiment" that laid the ground for the splitting of the atom. He famously declared: "Words or language . . . do not seem to play any role in my mechanism of thought . . . my elements of thought are . . . images."

As we progress I'll challenge your creativity by giving you problems to solve and will show you how using the mind's eye can help you come up with more novel answers. Word-free imagery is the surest way of escaping handicapping cliché and the leg-irons of mind-habit. I'll show you how logical, analytic thinking suits only certain types of problems. For more creative, intuitive-insightful thought, words can act as glue rather than grease in the cogwheels of thought. It is precisely these types of intuitive, creative thought processes that predict success in life better than standard, logical IQ-type tests—at least in people who are already above average in IQ. A famous study of world leaders showed that the higher their conventional IQ, the lower was their level of eminence as rated by independent experts.

In Chapter 6—"The Landscapes of Memory"—you'll see that you can use words or images, or both, when learning and remembering. Most people neglect the power of visualization when trying to learn, yet when you use both words and pictures to remember information you are using the two halves of your brain and hence learn better. This is particularly important for older people because visual memory holds up better with age than language-based memory, yet older people mostly do not use this brain potential to help preserve their memories. You can train yourself to greatly improve your memory by using imagery.

Chapter 7 is about stress and the mind's eye. Our most extreme emotions—fear, joy, desire, anger, despair—are all linked to powerful images we visualize. I'll show you how, untamed, these images can worsen your anxiety, but when used and controlled they can also rein in negative emotions very powerfully indeed. Visualizers may be more vulnerable than verbalizers to long-lasting stress after a trauma because the trauma lives on in their mind's eye, perpetuated by their visualizing power. But fears are also best tackled in the mind's eye, and you can use visualization to change how you feel and overcome your fears.

Our cravings and miniaddictions are also incubated in the

mind's eye. We visualize the sights, tastes, smells, sounds, and touch of what we crave and in so doing cook up a greater desire, reducing our resistance. The more easily you can visualize and absorb yourself in scenes or images, the more at risk you are of both allowing your fears to grow and strengthening your addictions. But you can also use the power of the mind's eye to help overcome addictions, through repeatedly imagining yourself reacting differently to the triggers that stimulate the craving.

We'll also see the part that visualization plays in health and immunity. Dramatic changes in your immune response—the ability of your body to fight disease—can become linked to particular triggers in your environment. In other words, your immune system can learn to weaken or strengthen according to the situation you are in. Take people undergoing chemotherapy for cancer: they can start to feel sick even at the sight, smell, or thought of the clinic where these nausea-inducing drugs are given. Visualizers' brains learn these kinds of links more readily than nonvisualizers'. But they can also learn to use imagery to overcome these problems and to help fight illness. Visual imagery can help treat skin conditions such as psoriasis—the mind's eye can alter how the cells in the body react. Imagery can also be used to help control several different types of illness, including migraine—even in children. And people given imagery training before major surgery recover better afterward and have less pain.

Imagery can greatly enhance athletic skills and strength through shaping brain circuits. In Chapter 9—"Visions of Olympus"—we will see how visualization is used by almost all the world's leading athletes. Tiger Woods was taught by his father to visualize the ball rolling into the hole as he hunched over concentrating on his putt. The people who are best at practicing their sport in their mind's eye tend to be the best achievers in anything from archery to tennis. You can even increase your physical strength purely by visualizing yourself doing the exercises.

Chapter 10 tackles the mysterious phenomenon of hypnosis. Recent neuroscience research has shown that hypnosis does indeed produce a change in brain functioning, particularly in the right half of the brain. If you "see" a nonexistent red apple under hypnotic suggestion, your brain will behave as if it is really seeing a red apple. Hypnosis can also reduce pain by changing your brain's response to the painful stimulus. Hypnosis relies heavily on the brain's capacity for imagery. The more vivid an imager you are, the better a subject of hypnosis you will tend to be.

What about the images that fill our dreams? This is a question for Chapter 11. There are two main kinds of dreams: those during rapid-eye-movement (REM) sleep, and those during non-REM sleep. In REM dreams, your capacity for imagery is unleashed because of the changed brain chemistry of sleep, and also because the brain's managers—the frontal lobes—are switched off.

Finally, in Chapter 12 we'll see how images are central to many religious practices and beliefs, and the key to some of the most profound experiences in our lives. Practices like meditation produce distinct changes in the brain corresponding to what people experience during these exercises. Different states in the brain can correspond to profoundly different types of consciousness.

Imagery-based thought, emotionally evocative and often creative if used sensibly, can help you discover new strengths and overcome old weaknesses. We need to cultivate a balance between logical, language-based thought on one hand, and intuitive, imagery-based thought on the other. That is what I hope this book will help you achieve.

2

The "Watery Clasp" of Language

It is 15,000 B.C.*, chill dawn in a glacial wilderness. A young Cro-Magnon man crouches in the snow-flecked heather, staring fixedly at a deer. It stands etched on the luminescent mist, wide-eyed, nostrils flared to Paleolithic man's rank scent.*

His smoke-blackened fingers dig unconsciously at the unyielding, frozen soil. Heart pounding, eyes flickering back and forth over the animal, his body is taut with their duet of mutual stillness. A mind full, quite full, of just this single image visualized at the crossroads of death and survival.

A hissing arc of birch and flint, the dull thwack, a scarlet, gorgeous spurt, and the stone-deflected scream as it rears and falls. It scrabbles for purchase in its own vivid and mist-suspended remnants, which in just one small corner of the conscious universe stay high and gold and vivid.

Half-crouched with the burden of this image, he makes his way down the stone-strewn slopes, leaving the others with their bloody loads. He crawls past the women's questioning eyes and deep, deep into the darkest spaces of the cave. His eyes burn with the strain of carrying it.

Hunched and cramped in this unfrequented corner, the wall glows in a slit of light exhausted by its long penetration through the dark. His eyes spill the resurrected deer onto the wall, and with the reverence of hunger, he traces with sharpened charcoal the tense, still lines of this projected image of its final earthly moment.

A Glimpse into the Paleolithic Mind?

In southwestern France and northern Spain, the present-day Basque people who live there are the direct descendants of a genetically distinct line of humanity. The Basques' direct ancestors may well be Cro-Magnon man, the Paleolithic authors of the earliest known artistic pictorial depictions by mankind.

In the deep, dark caves of Lascaux, Altamira, and other sites in these Basque regions, exquisitely painted and engraved outlines of deer, bison, and other animals appear with breathtaking lifelikeness. Beside them, though, matchstick humans prance awkwardly like the doodles of an infant. Why were these prehistoric artists so bad at drawing humans, but so good at drawing deer and bison?

Julian Jaynes of Princeton University suggested twenty years ago that these animal paintings weren't really "art." Rather, he argued, they might be a mechanical tracing of a vivid mental image projected by the eyes and brain of the draftsman onto the dim cave walls. This type of image—known as *eidetic imagery*—is present in as many as 1 in 10 present-day children, but hardly ever in the modern adult. It's a special kind of mental imagery, not properly understood, where a near-photographic image can be stored in the brain and projected onto a wall or screen like a slide.

Jaynes thought that a hungry hunter who had just sat still in the heather for two hours before dawn would be particularly prone to this kind of imagery. This would be especially true if after all those hours of cold and boredom his adrenaline started to pump at the sight of a deer emerging out of the gloom. The unique brain chemistry caused by hunger and excitement could conceivably release these eidetic imagery abilities and the pictures that followed. What's more, the cinema-like darkness of the deepest cave recesses gave pretty good conditions for the "projection" and copying of these mental images.

Imagery of all types seems to wither as our children grow up—including eidetic imagery. Is this because of our education system's focus on language? Let's take a look at the evidence.

Imprints on the Mind

Take another look at the picture of the horse at the beginning of this chapter. A four-year-old named Nadia drew this. Nadia was an autistic child with very poor language ability, the strange mannerisms and obsessions that are common in autism, and a highly abnormal brain. At times she would have uncontrolled episodes of screaming and destructiveness, alternating with periods of lethargy, withdrawal, and muteness. She was a clumsy child. What Nadia did have, however, was a narrow genius for drawing. After a single glance at a picture or figure, she could draw it with an almost Leonardo-type level of skill. Face pressed against the paper, her clumsiness temporarily suspended in her absorption, she would produce these wonderful drawings.

What was it like inside Nadia's universe: was it a narrow funnel of raw, undiluted images, uncontaminated by language? Maybe it was a coincidence that Nadia steadily lost her genius when she went to school at the age of seven. There, thanks to dedicated teaching, she learned how to spell out the world in words. Maybe Robert Graves's diagnosis is not correct—perhaps it was not the "watery clasp" of language that smothered these amazing images. But the bald fact is that when Nadia began to master the rudiments of language, her talent was extinguished.

Images tend to be destroyed by words—particularly eidetic images. To name the picture or the taste is to destroy it, often. Does Nadia give us a glimpse into an image-filled, wordless world that many children inhabit? Is it a glimpse into how it was for the Cro-Magnon draftsman, cramped in the dark, damp corners of his cave?

Nadia is not a unique example of so-called *savant* genius—a rare talent standing alone among limited general mental capacities. There are many examples in the world of such Rainman figures—often autistic—with amazing talents for imagery, both visual and musical.

The "Wide Glare" of the Child's Day

The older we get, the faster time seems to pass. This is partly because it's harder and harder to have an experience that's completely new. Our brains are constantly comparing what's happening *now* with memories of similar types of situations we've known in the past. This is a pretty useful survival tool, because it keeps us from getting nasty surprises and helps us learn from experience.

But there are—you've guessed it—costs. When you classify experiences like this, you begin to experience the *class* and not the event. In other words, your conscious experience becomes once removed from the immediate sensation. "Oh, I see the roses have come out" becomes a classification of the experience one step away from leisurely, wordlessly staring at a single rose. I say "leisurely" because it is a very automatic brain habit to classify experiences in this way, and you need time and effort to slip out of the classification mode into the wordless experience mode.

Here's an exercise: Lift your eyes from the page for a moment and look at some object around you—a chair, a cloud, a leaf, a face, a cup—anything. Look at it for a while and try to go beyond your normal "Oh, that's a chair" classifying response to it. Try to look at it as if it is unfamiliar, alien, and hence unclassifiable in words. Try not to describe it to yourself in words—rather, try to experience it through your senses. If you can, touch it, pick it up, and try also to avoid its classification through the sense of touch.

What classifying does is to make *categories* out of our experience. And the essence of categories is that they highlight the ways in which things or events are similar, and downplay the ways in which they are different. So the more we categorize our experience, the more day-to-day events will seem repetitive and similar: This is a recipe for time racing by in a blur.

Words play a big part in categorization, but it's harder to categorize when you bypass words and use your brain's capacity for imagery. In the hectic, information-flooded lives we live, we rely more and more on mental categories to filter the rush of data at the gateways of our experience. Eating an apple, for instance, has become for most of us a barely attended category of eating experience, rather than the multichannel tidal wave of sensation that it can be.

The word processor I am writing on just now works a bit like the human brain. If I start to use a phrase that it recognizes as one I have used often before, it guesses the rest and completes it for me, saving my fingers the work of typing it all. Sometimes it is wrong, more often it just doesn't quite hit the mark. But even then, sometimes rather than correcting it, I think, "Okay, computer, that'll do—have it your way." It's a kind of laziness, I suppose—but it's also efficient. After all, most of what we do, say, and think is routine. The tracks of habit and thought guide us around the swaying corners of life. Survival in the Western world would be impossible without these rails of routine.

Our brains are predictive machines also. They make moment-to-moment forecasts just as my word processor does. Read this sentence:

> *Fortunately for the drowning child, a woman on the beach had seen her and called the*
> *the rescue boat, which was soon on the scene.*

No doubt some of you will have seen the error in this sentence, but many of you—probably most—will not. If you didn't, have another look. Did you get it? There were two *the*s before *rescue.* The reason that many of you will not have noticed this is that your brains were acting just like my word processor—predicting what would be said, and seeing the prediction rather than the reality.

Each One of Us Is Blind

Imagine the following scene. You're walking across the lobby of a big hotel when someone you don't know comes up and asks you for directions. While you're giving the directions, two men pass between you and the stranger, carrying a door. You think this is a bit rude, but they move on and you continue describing the way the stranger has to go. When you've finished, he thanks you and then says, "You've just taken part in a psychology experiment. Did you notice anything change after the two men passed with the door?"

"No," you reply, puzzled. Then he tells you that he is not the man who originally asked for directions. That first man comes up to join you. You look at them side by side and they are completely different—different height, complexion, hair color, build, and dress. "You're joking," you say disbelievingly. "No, we're not. The first of us walked off behind the door, and the other slipped in in his place."

This experiment, led by Harvard psychologist Daniel Simons, showed that roughly 50 percent of people didn't notice that in the course of a couple of seconds, the stranger they were talking to was replaced by a completely different-looking person. How can this be? This *change blindness* is another example of how much of the time we don't really "see" the world around us. When the stranger comes up for directions, we tend to treat him as another category—here is a stranger and I have to work out how to tell him how to get to where he wants to go.

The key here is *attention*. We are attending to the instructions and not to the person. In fact, the person is irrelevant to the task in hand. What's more, in the jumble of experience that assails our eyes, we can't possibly take all of it in. Hence our brains tend to "fill in"—based on memory, stored images, and experience—this flotsam of background information. But if we tend to project old stored categories onto the world rather than actually seeing the full detail of the scenes in front of us, why don't we get knocked over by cars and buses, continually bump into tables, and ignore people we know when we see them in the street? Well, actually, people do all of these things from time to time, but for most of the time we manage to get around not too badly. This is because our brains are particularly sensitive to *changes* in scenes. So if the stranger you were talking to suddenly walked off and another replaced him, you would see the movements and other changes and would have no trouble noticing the impostor. But because the change happened behind the door the two men were carrying, at least half of the subjects didn't detect it.

Magicians are masters at using this change blindness. If a card is quickly swapped while your eye is moving from one position to another, then your brain probably won't notice the change. In other words, for the fraction of a second when your eyes are moving, you are effectively blind. Why then don't we experience the world as a sequence of flickering images interspersed by periods of blindness? We don't because our brains "fill in" the gaps and smooth out the world with remembered categories and rough sketches of experience.

There are other examples of this. The same Harvard researchers showed people a video of a basketball game and asked them to count the number of passes made by one of the teams. A minute or so into the match, a man in a gorilla suit walked slowly across the court, passing in among the players. Though clearly visible for about five seconds, again only half the viewers noticed him. Watching the same game again, but without any particular task to do—such as counting passes—they saw the gorilla easily, and found it hard to believe that this was the same video they had watched a few moments before. Again this shows that we miss much—indeed most—of what is in front of our eyes, ears, and other senses.

Wherever you are just now, pause for a few seconds. Make a note of all the different sounds you can hear. Don't give up after one or two: persist until you have a list of ten, twenty, or more. Make yourself aware of the orchestra of sounds that has been tickling your eardrum but never reaching your conscious mind. At this moment I am writing this in the lobby of a busy conference suite. I have been aware of the piped music intermittently while writing, and now the hum of the floor polisher comes to my attention. The *clip-clop* of heels on the polished floor, the hum of voices gradually becoming distinct individual voices as I attend to them . . . The grumble of a passing bus, the soft click of the keys on the keyboard, and a faint hissing from the ventilation vent just above my head, and indeed I become aware of the soft ringing of a faint tinnitus in my ears . . . and so on.

Taking the time to observe reveals layer upon layer of sensation that is being ignored. You can do this in any sense channel: take a mundane piece of bread, and really *attend* to the sensations in your mind as you eat it. You will become aware of gradients of texture and taste that you have probably always ignored through the thousands of pieces of bread you have eaten in your life. Try your own body at this moment—bring it into awareness and suddenly become conscious of the huge barrage of sensations that you have not been attending to.

In other words, we are not only blind and deaf to most of what is going on around us, but we are also oblivious to the sensations in other senses and in our own bodies. Nadia seemed to be nearer the minutiae of her visual world than you and I are. If we ask a four- or five-year-old child to draw a horse, and then compare it with Nadia's horse at the beginning of this chapter, will we see a difference? Is the drawing what the child actually sees? Or does it look as if the child was drawing categories of horseness downloaded from general descriptions stored in memory?

I failed art at school. My seven-year-old son draws better than I do. Yet at a conference recently, during a tedious lecture, I found myself sketching someone's face. And I accidentally discovered something that I now realize every real art teacher knows. I discovered that if you forget what it is you are drawing, and simply try to trace the lines and shades of the unnamed, uncategorized blob before your eyes, then you can get a likeness of sorts.

I am still no Leonardo, but that experience convinced me that to a considerable extent drawing is a skill that can be learned. Try it. But a precondition of learning that skill is shaking off the "cool web" of categorization. To draw you have to come much closer to the raw data of the senses, and switch off the machinery of naming and categorization.

The reverse may have happened to Nadia. Words are the foot soldiers of category. To name the lithe, furry catness before me as *cat* is to obliterate the particular with the general. So it is with the child visualizing the eidetic image of the cat: in naming it, the image shrivels to dust, and with it the unique particularities of *that* cat, destroyed by a category. Nadia could access uncluttered, uncategorized horseness until she learned to name it.

The brain's predilection for prediction and categorization is not confined to the visual sense. It also anticipates what we see, hear, feel, taste, and smell. Before the wine has even splashed across my tongue, my brain has unconsciously and automatically summoned from memory the taste I expect.

At a routine meal with your partner or friend, surreptitiously substitute for his or her drink a similar-looking but different beverage—dark cordial for red wine, white wine for grape juice, tea for coffee, etc. Watch for the reaction when his or her brain's prediction is confounded. Ask what it tasted like. Did it taste like either the expected or the actual drink? In most cases, the response will be that it tasted strange and foreign—not like either.

Wine will not taste like wine for that second that your brain has prepared itself for something else. For much of our lives, we taste memories—what we expect—not the raw, fresh complexity of the sensations on our tongues.

It cannot really be any other way. The billions of bits of information that a single scene might contain—texture, shape, shadows, objects, movement, location—could never be simply transcripted "raw" into our brains via the senses. Of course we must impose windows through which they are channeled. And the most important of these windows is attention. What we attend to has a royal road into our brain circuits, though there is much that we don't attend to—and of which we are unaware—that also imprints itself on our neural circuits.

But the outside world can hijack our attention. Overly loud music in a restaurant can drag our attention away from the subtleties of the food. The comically nodding toupee of a lecturer can obliterate the words of his excellent speech. Here the unexpected, the comic, the harsh, the frightening, the sexy, and the emotional romp into our consciousness like a rampage of soccer hooligans. It is perhaps at these rare moments that we are closest to the unfettered, uncategorized *seeing* that we attribute to young children and to savants.

Great comedians and artists are loved and remembered because they help us see what we take for granted in a new—often absurd—light. Their genius burrows through the "seen that, done that" habits with which we usually perceive the world. But it's becoming harder and harder for them, for less and less is new. We even categorize jokes, and so neuter them. The art world is becoming almost desperate in this struggle to break through the clichés and habits of perception. We find a whole movement in visual art that tries to break through the barriers of perception by jarring us with images that provoke disgust. This is not necessarily the art of cheap thrills. If art is about anything, it is about helping people see the object and not the category.

Maybe—just maybe—young children really are closer to the "black wastes" and the "summer rose" of Robert Graves's poem. Perhaps this is why time is endless for them, for how do we trace the passage of time but by events? And so long as these events and experiences are fresh and uncategorized, time must surely slow to accommodate them. As we get older, events become categories of experience—generalities and replications of some distant fresh experience of the senses. One plane trip merges into another because of how our brains have learned to code them: "we took off, we landed," not the first awesome sensation of your body lifted into thin air in a juddering tube of metal. Children cry or shout at the "wide glare" of the looming sky. We name it and obliterate the awe.

Nadia can't describe hers any more than Paleolithic man can come back to tell us about his. Still, perhaps there is someone who can hint what it might be like—a remarkable Russian journalist who lived during Stalin's time.

Seared in the "Wide Glare"

What if we didn't categorize? What if we lived standing at the glass doors of perception, exposed to the full daylight glare of sensation? Would we, as Graves's poem suggests, "go mad . . . and die that way?" A Russian journalist who seemed to live near this state suggests that we could survive. But we would be as disabled by the particular as we can be by the category.

The great Russian neuropsychologist A. R. Luria wrote beautifully and meticulously about this journalist, S, whom he studied over many years, in his book *The Mind of a Mnemonist*. Though the title of this book refers to S's memory, it is as much about the man's minute-to-minute encounters with the raw data of sensation.

S had a mind-boggling memory, and ended up living his

tenuous life as a stage mnemonist—demonstrating prodigious feats of memory for a fee. In the 1920s, however, he was a young reporter, and it was his memory that had caused his editor to send him to Luria. While other journalists would take copious notes at briefings and interviews, S needed none—he remembered almost everything in an uncannily raw form. But Luria soon found that the basis for the journalist's memory was in his powers of *imagery*. And this was not just visual imagery—his was a multimedia carnival of taste, touch, sound, and smell as well. He experienced words as puffs of steam, for instance. A certain noise had the taste of sweet and sour borscht, a sensation that gripped his entire tongue. He felt a high-pitched tone as a needle stabbed into his spine. Once, when he heard a bell ringing, he saw a small, round object roll before his eyes, his fingers sensed something tough, like rope, and he suddenly had the taste of salt water in his mouth. All this from a simple bell ringing!

S had the power of synesthesia—the ability to experience sensation in one sense modality through another. His memory was based largely on his visual imagery, however, he had an incredible power of eidetic imagery. He would remember a long series of words by laying them out in various spots along a mental road where he took an imaginary walk—a word by this lamppost here, another in that corner there. When he forgot these words—which he rarely did, even over the course of many years—it would be because he had laid the word in an unlit corner, for instance. When he took a later walk down the street to "pick up" the words, he "didn't notice" a particular word because it was in a dark corner. But when he learned lists of words, S would hardly ever see patterns in their meaning. So, for instance, if he had to learn *dog, swing, sky, cat, cow, horse, chair, door, bed,* he would never categorize the words into "animals" or "furniture." He would be so caught up in the multisensory sensations the words evoked that he would fail even to see this simple pattern.

You and I rely precisely on this type of categorization for memory—without it we would remember much less than we do. But S's was an entirely different way of remembering—and perceiving—the world. S had escaped the "cool web," and lived in the "wide glare of the child's day." You might think that S would have had a rather successful life, given these prodigious abilities. On the contrary, he led a rather disorganized, even feckless life. And this was in part because he seemed unable to transcend the particular and learn to categorize and generalize in the way that we all need to do to survive in an industrial—now digital—economy.

When S read prose—fictional or nonfictional—he had great difficulty extracting meanings and concepts because he was so distracted by the multitude of images and sensations that single words would trigger—puffs of steam and splashes, for instance. For a person so tied to the visual image, abstract concepts that were hard to visualize were a real problem: he couldn't grasp the concept of 'infinity,' for instance. To Luria, he even came across as rather dull-witted. He had difficulty understanding Luria's stories, because the words called up images that collided with each other, resulting in chaos. He couldn't even ignore the quality of sound of Luria's voice, which set off multisensory images in his mind.

Is this something like the way it is for young children? We know that even five-year-old children who have a good command of language tend to think more in images than in words. By the time they are ten, with five years of schooling under their belt, however, they are word- and sound-focused and do not use images to the same extent.

The work of Graham Hitch in Lancaster, England, proved this. He showed children pictures of objects—pens, knives, umbrellas, kangaroos, etc. As they looked at them, the children could concentrate on the name of the picture or its visual image—it was left up to them. Those who had progressed well in the school system were probably more likely to use words. And if they used words, then long words like *kangaroo* would fill up the short-term word memory system in the left hemisphere of their brains—the so-called phonological loop.

The phonological loop is the mind's system for holding on to sounds—usually words or numbers—for the few seconds we need to use or check them. "Did he really say that . . . ?" This brain system allows you to replay sounds you have just heard but not to store them in memory for any longer than it takes to do just that. You need the phonological loop for remembering the telephone number that someone gives you during the few seconds it takes to write it down. And—most importantly perhaps—you need the phonological loop as a child to store the sounds that make up these long words you have to learn to read and pronounce—*hip-po-pot-a-mus.*

Words such as *hippopotamus* or, less exotically, *umbrella* will be harder to keep in mind because they have more syllables and use up more of the limited space in the phonological loop than short words do; if children are relying on this wordy loop, then they should make more mistakes. Short words like *pen* and *knife,* for instance, will be a piece of cake for children using the phonological loop because they sound quite different and won't overtax the loop. If children are using the mind's eye to take in the objects as *images*, on the other hand, then it won't matter whether the names are long or short—the mind's eye doesn't care about the sounds, just about the visual appearance of the objects. But this visually based memory system—very likely located in the right hemisphere of the brain—*will* have problems in mixing up objects that *look* the same: *umbrella* and *cane,* for instance.

The results of this study are easy to predict. If we are right in saying that young children haven't yet learned to spell out the world, and think in images rather than in the abstract code of language, then they should show more errors when they are remembering objects that look the same, and fewer when the names of the objects are long. The converse should be true for the older children whom we presume to be—after five years in school—wrapped in the "cool web of language."

This is exactly what Graham Hitch and his colleagues found. Five-year-olds clearly used mental images much more when looking at the pictures, and hence were not penalized by pictures that had long names. The ten-year-olds, on the other hand, didn't make as many errors when the objects looked the same, but had problems when they had long names. Robert Graves, we see yet again, was on the right track.

We Are All Amnesiacs

What are your earliest memories? Do you remember your third birthday? If you have one, do you remember when your younger brother or sister was born? What happened on your first day at nursery school, if you went to one? What about your first day at school? Take a few moments to try to remember. You may want to write your memories down.

I remember a cot in a bedroom with a gas fire. I remember a quilt with squares—blue, with little figures on it. I remember my father coming in with an orange, and I remember pulling off the white bits from the orange. I remember the smell of oil and factory from his overcoat.

That's one of my first memories. There are others, but it's hard to know how old I was. Given that I was still in a crib then, probably two or three. The average adult can remember one experience from the age of two to three. The figure for age three to four is three, four to five is six, and five to six is ten. It is very rare to remember anything from earlier years.

The Russian S had quite detailed memories from his early years, and these memories had a preverbal quality to them that, taken with the events he describes, makes it possible that he was remembering from a time earlier than age two. He remembers a cradle, the discomfort of what might have been his diaper being changed, being rocked, crying, and then crying more at the sound of his own crying. He remembers being sat on the potty and being frightened of it.

Of course we can't tell whether or not these were real memories from his prewalking days. He could well have been imagining these things. But there is a spellbinding persuasiveness to his descriptions of a wordless world of sensation that makes you want to believe that there was someone who really hadn't erased these memories. What seems clear at least is that in the preverbal world of the infant, image is king. That's why the memories we have of infancy are like flashbacks—wordless sensations, visions of quite vivid quality, glimpses of the "wide glare" of the child's day.

But why are we amnesiac for these early years? Psychologists used to think that the brain systems responsible for a particular kind of memory—episodic memory—hadn't developed properly, but recent research casts doubt on this. *Episodic memory* is the name for a memory of a certain event in a particular context. For instance, remember the last time you watched a movie. When you recall it, you won't just remember the movie—you will also retrieve where you were, whom you were with, and many other details. Contrast this with *semantic memory*—your stored knowledge of bald, contextless facts. Recall, for instance, the name of the capital of Russia—Moscow. You know that Moscow is Russia's capital, but it's very unlikely that you'll remember where you first learned this fact. Semantic memory depends on a quite different brain system of memory from episodic memory.

Many scientists believed that infants' brains simply hadn't developed enough for them to have episodic memories. Episodic memory needs the frontal lobes of the brain, and these are the last parts of the brain to develop. On the other hand, semantic memory—remembering that dogs have four legs and wag their tails—was thought to depend on brain systems that developed much earlier, so that infants could have semantic memory but not episodic. Not so, it seems. Infants as young as thirteen months can recall specific episodes eight months later. Patricia Bauer of the University of Minnesota showed that infants acted out stories using sets of objects on the table in front of them. For example, the child sees a frame, a rod, a hooked metal plate, and a mallet. He watches the psychologist put the rod on the frame, hang the metal plate on it, and hit it with the mallet, saying, "Let's make a gong," talking the child through the actions as she goes. Eight months later, the children see the same objects again, and—more likely than not—they show that they remember the episode by "replaying" some or all of the specific actions they witnessed so long ago.

Why is it, then, that if infants can recall episodes like that, we can recall so little of our infancy as adults? The increasing importance of language's "watery clasp" as we grow older offers one explanation. It may be that these early memories aren't encrypted into verbal code at the time, and so when we try to recall them years later with our predominantly verbal apparatus, we simply can't access them. It's as if the message is written and sent using one codebook, but received by people using a quite different code. In a sense, as an adult you have lost the codebook of wordless image, and so the image-encrypted messages of infancy are inaccessible to you.

Perhaps our journalist from Stalin's time was able to access the images and memories of infancy because his mind's eye had little succumbed to the clutches of language. But is there any evidence to support such a view? Yes, there is.

When infants are tested for their memory of acted-out scenes, they also tend to make little comments along the way. "Hang the plate on the bar . . . Hit the gong." This reveals two ways of measuring infants' memory: what they say . . . and what they do. A child could *show* that she remembered by doing what she did before. Alternatively a child might *say* things about the previous visit but might not demonstrate his memory by doing what he did before. Patricia Bauer watched how well her various infants could speak and understand. She was particularly interested in how well they could use language the first time they saw the scenes they were later to remember. You might assume that the children with the better developed language would be the ones who would remember best—by showing—what had happened months earlier. Not so. The wordy thirteen-month-old children were actually quite poor at showing they remembered—judged not by what they *said* but by what they *did* when faced with the rod, the gong, and other objects.

What the child's language did predict, however, was their memory measured by what they said about that experience eight months before. In other words, children could show using action and image that they remembered what had happened months before, but this memory was inaccessible to them in words. There were two different codebooks and two different codes—one for words and one for images.

So we have some scientific support for the notion that word and image have somewhat separate lives within our skulls, and that probing the experience of one from the domain of another can be difficult. One reason why we remember so little of our infancy may be because early experiences tend to be written using the codebook of image, but we try to remember them using the quite incompatible codebook of language. S may be a near-unique example of an adult who could decipher both codes.

When my son was three, he went on a nursery-school trip by train to a children's farm and zoo near Cambridge, England. At the station on the journey back, the train conductor carelessly let the train doors shut, leaving him and—fortunately—one teacher on the platform while the train pulled off with twenty other children and three teachers. This was a pretty major event in the life of a three-year-old, and when I asked him today—four years later—about whether and what he remembers of the event, he told me he remembered the teacher buying him ice cream and the train doors closing. He reported these events in reverse chronological order, and his storyline was similarly quite weak.

Another class of nursery-school children experienced a comparable earthshaking event—evacuation of their school following a minor fire. This time, however, some psychologists saw the opportunity to study what these children would remember years later of the event, and how. What they found was a considerable difference in the coherence of the memories of this event seven years on, depending on how old the child was at the time it happened. The three-and-a-half-year-olds, on average, remembered only isolated images from the event—just as my son did. The four-and-a-half-year-olds, on the other hand, were much more likely to have an orderly, structured narrative recall of the events where cause was succeeded by effect and time 2 followed time 1. What had happened to the brains of the four-year-olds that allowed them, seven years later, to recall the events so much more coherently?

Well, for one thing their language abilities would have mushroomed in that year, and if narrative is based on anything, it is language. *Then, because, after, before, why*—these are the scaffolding of narrative; they are also part of the abstract code of language and very difficult to replicate in the world of the mind's eye.

Our memories, then, are stories? In a way, they are—but stories studded with images that illustrate the narrative. Without the words we are left with isolated visions—often emotional, colorful, and vivid, but nevertheless as fragmented and confusing as dreams if they are starved of the narrative power of language.

Do we remember so little of our early childhood because we don't know how to make up stories for ourselves? Yes, but there is more to it. Stories need actors, participants, protagonists—call them what you will. And who is the main character in the stories I construct in my memories? You've guessed it. *Me.*

The bottom line of this theory of childhood amnesia proposed by psychologist Josef Perner is this: To truly remember something that happened, in the sense of having the sense of reliving that event, one crucial ingredient is required. And that ingredient is a form of self-consciousness: When the event is occurring, the child must register that he or she is experiencing what is happening in a self-knowing and self-conscious way. "This is me experiencing this"—not necessarily verbalized, but nevertheless known by the child at the moment of experiencing—becomes the crucial part of the context that buys the episode a ticket into the story of self that we call "my memory." In other words, according to this theory it is the very unself-consciousness of the toddler that is the source of his or her amnesia.

And that unself-consciousness exists in a largely preverbal glare of image and emotion where events are wide and awful, sensuous and inarticulate—just as Robert Graves's poem describes them. Only when I begin to tell to myself these moments of raw experience as a story of *then* and *now* and *because* and *when* do they harden into the facts of my autobiography that I can recall from my tent on the blown steppes of adulthood. Words are the battered trunks that carry remnants of experience across the plains of time.

Endel Tulving—one of the greatest living figures in psychology—first made the distinction between episodic and semantic memory, and he too has forced us to understand that a sense of self is crucial for memory of a certain type. We call this memory *autobiographical memory*—the collection of memories that we treasure as intimately and uniquely ours. These episodes, where *I* is omnipresent as the lynchpin of both the remembered and the remembering, are central not only to our personal memories but to our very core of self. Young children are blessed—and also cursed, if you are to accept Graves—by a freedom from this watchful *I* registering that "this is something I am experiencing." So when they come to try to recall events from the pre-*I* years of infancy, the key reminder that will jog the memory is missing.

Except, that is, when fragments of a preverbal world of image tumble past the sleeping sentries of narrative in dreams and in other altered states of consciousness where the codebook of language is temporarily closed.

So we can remember a lot more than we think of these early years, can we? Probably. But we can't distinguish fact from fantasy in these early memories. Young children can remember things they have seen and they can remember things they have been told. The problem is, they find it hard to tell the difference. What is needed here is *source memory*. It means what it says—memory not for *what* you know, but where and when you learned it. Source memory is a skill we owe to our frontal lobes. It is a critical element in the billions of items of experience we accumulate every day. Sensations and experiences are culled ruthlessly by attention—another feature of the frontal lobes—and sorted out according to the context. Context consists of the other sensations, memories, and thoughts that happened to be around at the time, and to which memories are stuck like flies to a windshield. And the windshield is, if we are to accept Tulving, that story we call *me*.

Me is not a major player in the experience of young children, and so the flies of memory don't have anything to stick to. The flies are still there—the memories are there—but young children tend not to know how they got there. In one study, for instance, children of four, six, and eight years old were either shown events or asked to imagine them. For instance, they might actually unpack a picnic basket, or they might just listen to a tape asking them to imagine this.

Tested a week later, they were asked questions about the various events they had either participated in or imagined; for instance, what color the napkin was in the picnic basket, or what shape the basket was. Six- and eight-year-olds remembered more than the four-year-olds, not surprisingly, though they were far from perfect. But what the six- and eight-year-olds *could* do very well was say whether the event had actually happened to them, or whether they had simply imagined it while listening to a tape. The four-year-olds, on the other hand, were very poor at being able to distinguish the source of what they remembered: they found it hard to know whether it had happened to them or whether they had imagined it.

So while the preverbal images of childhood may be vivid and immediate, they are not reliable markers of reality. On the contrary, the source of their power and freshness of vision—as we see, for instance, in Nadia's horse—may be their location at the crossroads of fantasy and reality. But what *is* this power of imagery? How do our brains construct these wordless pictures in the mind's eye? These are the questions for the next chapter.

3

How Your Brain Creates Images

He works the wet clay, molding it to the shape of her. Out of this inert gray substance, the vital, curving geometry of her body swells in an alchemy of life out of earth. He turns it on the sculptor's turntable, and the model's eyes follow the rotation of her re-created body. Turned full circle, he tugs and presses at the living clay some more, nursing it further into life.

Then he steps back, the artist's gnawing discontent at least appeased, if not satisfied. He swivels her sculpture for the model to see, and is surprised to see her looking puzzled.

"It's . . . beautiful—like it's alive," she says, "except . . . except—what happened to my body there?"

"Where?" He frowns and looks at it closely.

"There."

"What do you mean?" he snaps. She looks flustered.

"I'm sorry—maybe I don't understand—maybe it's meant to be like that."

"Like what?" He's looking angrily from her to the sculpture and back again.

"Like . . . like I'm all sort of misshapen . . . on one side. Look, my arm just sort of fades away."

He's looking intently at the sculpture, scanning where her finger is pointing, frowning as if he has an inkling that something is not right but can't quite work out what. He stares at his sculpture, at the flawless shape of the right side of her body—the left side from his perspective. He moves it to the right and left a few degrees on its turntable. Something isn't right, he knows now, but it's so hard to put his finger on precisely what. He takes off his glasses, cleans them, and puts them on again. But no, he still doesn't get it. He doesn't get the gradual decay from curving, flawless right to deformed, disintegrating left of the clay body that he has created.

Most of Milan's citizens know the Cathedral Square in Milan—Piazza del Duomo—intimately. They have a detailed knowledge of the fine palazzos that mark out the perimeter of the square. At the University Neurology Clinic of the University of Milan, close by the Piazza del Duomo, Professor Eduardo Bisiach and Dr. Carlo Luzzatti are interviewing one of their patients, a distinguished Milanese lawyer who has suffered a stroke affecting the lower part of the parietal lobe in the right hemisphere of his brain.

"We would like you to close your eyes and picture in your mind the Piazza del Duomo. Imagine that you are standing facing the front door of the Cathedral. Can you see it?"

"Yes, I can see it in my mind."

"Can you see the great front door?"

"Yes."

"Can you see all the towers on the facade?"

"Yes, I have the picture clearly in my mind."

"Can you see the other buildings and features of the square?"

"Yes, I can see them."

"Okay, now tell me what you can see."

"I can see the corner of the Royal Palace, the Arengario, the Vittorio Emanuele II monument, the arcades, Galtrucco, Piazza Missori, Loggia dei Mercanti. . . ."

The patient goes on to list many of the features of the square. But there is something strange about his answer. The majority of buildings and streets he reports are on the right side of the square.

"Now we would like you to turn around in your mind. Pretend you are standing on the front steps of the Duomo, looking back. Can you see the square now from this perspective?"

"Yes, I can see quite clearly."

"Tell us the buildings you can see."

"I can see the palace with the arcade to Via Orefici, the Palazzo dei Giureconsulti, Motta, Piazza San Fedele, Rinascente, Via San Raffaele. . . ."

Again it is strange, for he describes the buildings, streets, and shops largely on the right side of the square again. But, of course, this is the old left side—the palazzos and other buildings that he didn't "see" in his mind's eye when facing toward the cathedral.

The Italian neuropsychologists interviewed a second patient. He showed the same pattern of distorted imagery in their mind's eyes—always missing the buildings on the left side of their mental image, even though the "lost" palazzos appeared again when they mentally turned around to face the opposite direction.

A Whole World Neglected

Young children think using images more than words, but as they learn to speak and read, their ability to think in images gradually withers. As adults, most of us have remnants of this capacity for imagery—but what *is* it? What is happening in our brains as we imagine in this wordless way? In this chapter I'll show you how different types of imagery are created in the brain.

The clearest way of seeing how the brain builds images is to study people who have suffered damage to parts of the brain that control imagery. The people described in the introduction to this chapter had both suffered strokes in the right half of their brains—in and around the parietal lobes. In Milan, Professor Eduardo Bisiach and Dr. Carlo Luzzatti had just carried out one of the most famous neuropsychological studies ever. Their patients had a bizarre but nevertheless quite common disorder called *unilateral neglect.* This means what it says—a "neglect" of one whole half of space, usually on the left side.

Neglect is not a question of blindness. The sculptor—this description is also based on a real case—was able to "see" the left side of his work of art, at least after his model had drawn attention to the decaying side of her clay body. Similarly, several of the Milan patients were able to see on the left if their attention was drawn to it, albeit with difficulty. What none of these people could do was to create images on the left side of their mind's eye. They were fine on the right—brilliant, in the case of the sculptor—but the left side of their mental screen had disappeared.

It's hard to put yourself in their shoes: How can you lose the image of half the world? It's much easier to appreciate what it's like to lose the power to understand language—after all, we face that every time we go to a foreign country whose language we don't speak. Imagining what it's like to have unilateral neglect is much harder than putting yourself into the shoes of someone who has lost the power of language, partly because we think so much in words. Neglect, however, is a malfunction of that other realm—the word-free realm of space and image that is the topic of this book.

Two Brains in One Skull

When you close your eyes and mentally picture a sight or sound, what are you doing? You are switching on several different clusters of neurons in different parts of your brain. When the links between these clusters are cut, the distinct worlds of word and image are unmasked. Let's see how this happens.

If you are unlucky enough to suffer from one of certain types of epilepsy that don't respond to normal drugs, surgeons may have to separate the two halves of your brain to bring it under control. This leaves each half of the brain isolated from the other half, because the connecting fibers—the corpus callosum—are cut.

When the first split-brain operations were carried out in California in the early 1960s, Roger Sperry—who in 1981 was awarded the Nobel Prize for Medicine for this work—and Michael Gazzaniga studied the patients. One of their tests had the patients look at a screen and say when lights flashed to the left and to the right. Because they were looking straight ahead, and because of the eye-brain wiring, the lights on the left side were seen only by the patients' right brains, and vice versa. When a light flashed on the right, they said they saw it. But when the light appeared on the left, they said nothing. Was it because they were blind on that side? No. Because when they had to *point* to whichever light came on, they pointed to the lights on the left, so they did see them. What was happening? Simple. The speech and main language centers are in the left half of the brain, so when the right-sided light is detected there, the information can pass down to these language centers to give the spoken response "I saw the light on the right." But when the left-sided light is detected in the isolated and word-deprived right half of the brain, the information can't pass over the gulf of severed fibers between the two halves of the brain to reach the speech centers. It *can* pass to the movement-control centers of the mute right hemisphere and make the person point to the light.

The experience of seeing a light on the left side of the screen was trapped in the inarticulate right half of the brain, and if the researchers had simply kept on asking the patients what they saw on the left, they might have concluded that there was no awareness in the right half of the brain. But the fact that the isolated right side of the brain could prove it had awareness by pointing shows that not all our experience and thought are captured or capturable in words and language. This is rather similar to what happened to the infants in the last chapter—they could show they remembered by what they *did* but not by what they *said,* and vice versa. In other words, it seems that these two brain systems can't read each other's codebooks.

Take a handful of coins, and pull out at least two of each value. Now create two equal piles on the table in front of you, one to your left and one to your right. Close your eyes, pick up a coin with your right hand, and try to recognize what it is. Now, still with eyes closed, and with your left hand, try to pick out the coin from the left-hand pile that matches the one you're holding in your right. This is not easy when the coins are similar in size, but you should be able to do it with some degree of accuracy. Your brain manages this feat only because of the thick wad of connecting fibers that passes the information from your left brain (right hand) over to your right brain (left hand) so that the two sets of touch sensations can be compared.

If you were a split-brain patient, then you might recognize coins you touched with your left hand but without being able to put that experience into words. The split-brain patients weren't able to name objects they felt with their left hands because that information went to the right side of the brain and couldn't then pass over to the speech centers in the left brain.

But when a picture of an object—a fork, for instance—was flashed to the right hemisphere, they were able to feel around with their left hand to pick from a selection of objects and find by touch alone the one they had seen visually, the fork. This even happened when, for instance, the right hemisphere was shown a picture of a cigarette, but the only related object for their left hand to find was an ashtray. They managed to choose the ashtray, even though they could not say what it was, or describe it. In other words, the right hemisphere "knew" something about the object, but could not express that thought verbally.

In another experiment, the California researchers occasionally flashed a picture of a nude woman to a subject's right brain, in among a series of ordinary objects. When they asked the subject immediately afterward what she had seen, she said that she had seen nothing. But then a smile spread over her face and she began to chuckle. Asked to explain, she laughed, saying something like, "Oh, it's just that silly machine." In other words, her right brain not only took in the picture, but also responded emotionally to it. All it could not do was articulate this experience in words.

If we want to build up our ability to create wordless images, then it's important to understand the different types of imagery that exist, with their corresponding different brain networks. When we create a visual image, we don't just metaphorically paint a static mental picture. We also move the image mentally in our mind's eye. The next exercise gets you to use this image-moving apparatus inside your skull. This exercise will particularly increase the activity of sets of neurons in the right half of your brain, behind and above your right ear in the parietal lobe.

Look at the rotated number at the beginning of the chapter. Is it a real 2, or is it a mirror image of a 2? To answer this question, it is almost certain that you will have to mentally turn the number to its upright position first. Do that now, if you haven't already.

Get some pieces of paper, and write one of each of the following letters and numbers on each slip of paper: F P G R 2 4 5 7. Now do the same again, but this time print the characters in their mirror-reversed form. Shuffle the two sets together. Close your eyes and scatter them over the table, so that they fall at all angles and orientations. Pick up one paper and glance at it at whatever angle it happens to be in your hand. Is the character normal or reversed? If you really glance at them at a range of different angles, you will find that you have to mentally turn them to an upright position in your mind's eye, to "see" if they are mirror-reversed or not.

The disconnected right brains of split-brain patients also do better at this kind of task than their left-brain partners. Neuroscientists now understand better how our brains—particularly the right half—manage to do this. You might have noticed that some of the numbers you rotated had to be turned farther than others in order to get them to their upright position. The greater the angle that the number is rotated on the page, the longer it takes for your brain to check the number in its upright position using mental rotation.

The same is true if you have to compare rotated abstract objects to see whether they are the same or not. This thirty-year-old discovery suggested that the image "moves" in the brain at a similar pace as it would if it were rotated in real life. Recording from single cells in the brain during the mental rotation needed to plan movement has suggested one way that this may happen. Cells in one of the movement-control parts of the brain—the motor cortex—are "tuned" to particular directions of movement. In other words, some cells are active for a movement in one direction, while others are active in a different direction. If you take an average "tuning" for a group of cells, it will have a generally preferred direction of movement, based on the average of all the cells in that group.

Think of it as similar to throwing a handful of coins: there will not only be a general direction for the coins as a whole as you throw them, but a wide spread as well. Nevertheless, if you took the average of the angles of all the coins, you would come up with a single average path—let's call this the average "tuning of direction" of the scattered coins. When you have to plan a movement to a mentally rotated point in space, cells in the brain's movement center steadily change their average tuning of direction in line with the mental rotation. Imagine that the coins were ferrous, and that as you threw them a large magnet was switched on that pulled the scatter of coins toward it. There would still be a scatter, but the average direction of the coins would have changed. During mental rotation, the average direction of tuning of the group of cells changes.

When you moved the letter around in your mind, it may be that your brain managed to do so by steadily swiveling around the average tuning direction of a group of cells—possibly in your right posterior parietal lobe. This would explain why the time it takes for the mental rotation is so closely related to how far the letter is turned around on the real page. A small angle would need only a small change in the average direction to which the group of neurons was tuned, whereas a big angle would demand a major shift in the average direction of the neurons.

This way of representing the outside world inside our heads is quite different from how we represent it using the "cool web of language." Taking the topsy-turvy number at the beginning of this chapter, you could have said something like, "That number is rotated 45 degrees and needs to be turned back by 45 degrees to get it to the upright position." But this would be a completely different, pretty inefficient, and wordy way of representing the world for this particular task. What you really needed to do was *visualize* the numbers, and the brain's visualization machinery is what this book is all about.

Where Did I Park the Car?

Though we tend to neglect imagery in favor of words, we can't do entirely without it. Take the example of the supermarket parking lot. You come out of the supermarket with the groceries, car keys in hand. You stop short. Where on earth did you leave the car? Now, if you had been paying attention when you locked it up—which you weren't, because you were in a rush—you would have noted that it was in row G, next to the wall beside the gas station. That would have been the wordy way to do it.

An alternative way would have been to make a wordless mental map of the lot. On that map would be a few key landmarks—supermarket entrance, parking lot exit, your car, for instance. These landmarks would be marked on your mental map so that when you came out of the supermarket, you wouldn't need to do much more than to dig this map out of memory, take a fraction of a second to scan it, and then march in the direction of the car.

Most of the time, most people probably use a mixture of word and image to find their cars in a parking lot. But you can probably think of people who fall at the extremes of these two dimensions. Some give verbose descriptions of their lack of sense of direction, while others are tongue-tied but perfect navigators. Between these extremes, we all differ in the balance of abilities in these two realms.

My colleague Martin Farrell of the University of Manchester

and I have studied how we manage to do this kind of imagery task—let's call it mental map reading. We asked volunteers to sit blindfolded in a swivel chair, surrounded by various objects on stands—a shoe, a cup, etc. Think of these as comparable to the car, the supermarket door, the parking lot exit, etc. They had already memorized what and where these objects were, so that when asked to point to the shoe they would point in its direction without difficulty. In other words, they had drawn a mental map of where they were in relation to these landmarks. So let's say this is you, just out of your car, who has glanced around and noted where the landmarks of the parking lot are. It may be that you didn't need to learn them because you have been there so often; you only needed to make a cross on the well-established mental map marking where your car is this time.

We then swiveled the chair around and asked the blindfolded volunteers to point to various objects from their new position. Think of this as like you deviating from your path to the supermarket entrance and walking over to read a sign on the wall. What was remarkable about the results was how easily our subjects pointed to the targets, even though they had been rotated to a new position. It was as if the map was effortlessly redrawn to the perspective of their new position, without much effort from them.

Maybe it's not so remarkable. You don't lose your car in the lot just because you've changed direction and walked over to greet a friend. Unless, that is, you haven't made the map in the first place, which is not uncommon in busy shoppers trying to do ten things at once. But, assuming that you have taken the fraction of a second it takes to locate the landmarks in the parking lot, then changing direction should still allow you to pinpoint your car pretty quickly, even from a new position. And how far you swivel doesn't—if you think about it—seem to affect the difficulty of finding your car from your new location. And that's exactly what Martin Farrell and I found: unlike the mental number rotation exercise you did earlier, the angle through which our volunteers turned didn't affect how quickly they could point to a landmark from a new direction.

Why this difference between rotation on the mental screen and physical rotation in real life? When you physically turn, the vestibular system just below your ears detects the speed and direction of movement, just as it does when you are seasick. It seems that this information is fired up to the map department of the brain, saying something like "Stop press: recalibrate all locations by 30 degrees." According to our data, this may happen automatically, without any effort on your—the busy shopper's—part. This is why our volunteers took no longer to point to the shoe when they were rotated by 180 degrees than when they swiveled through 45 degrees.

So navigating in a parking lot-type space is easy? Not always. Your cell phone rings just as you are entering the supermarket. It's your partner, who wants to leave a heavy bag in the car and will be coming in the side entrance of the lot. Can you describe where the car is from that perspective? We asked our volunteers to do something similar. We said to them, "Imagine you are facing the cup, and from that perspective, please point to the shoe." This is exactly what your partner is asking of you—imagine you are coming from the side entrance, and point to the car from this imagined point of view.

What we found was far more similar to the rotating-number results. The more our volunteers had to mentally swivel their chair, the longer it took them to point to where the shoe was from the new, imaginary position facing the cup. In other words, without the help of the vestibular system detecting real turning, you have to put in the effort of rotating yourself in mental space. And in some respects, mental space is a bit like real space, with greater distances needing greater real—and imagined—legwork to get there. As you stand in the supermarket trying to give directions over your cell phone to your laden-down partner, it's a far more demanding job for your brain, which has to move you around in mental space until you are facing the side entrance, and the car now has to be located from this new perspective.

Where in the brain is this happening? We studied people who had had strokes in the left or right sides of their brains. We found that the patients who had suffered strokes affecting the right half of their brains, in the parietal lobe, had the most difficulty. This is, of course, exactly the area that you need to do the rotation of the numbers at the beginning of this chapter. With these people, their brains did not seem to update their positions automatically when they moved.

Not that you have to have a stroke to have difficulties with this type of situation. We have found apparently perfectly normal people working in good jobs who get lost very easily and simply don't have a good sense of direction. It is likely that these people don't have an efficient parietal-lobe spatial system in their right brains, for whatever reason—genetic, prenatal, perinatal, or just plain random individual differences. They can function perfectly normally, but they clearly won't be architects, sculptors, or engineers, who have to be able to think wordlessly about space, shape, and location. No, people who have trouble with this kind of visualization of space will more likely be journalists, administrators, and lawyers, whose currency of thought is language. These wordy professionals won't necessarily be poor in all the functions of the mind's eye, however. Being able to think about space relies on a quite different brain network from other types of imagery. So if you are such a spatially challenged person, don't write yourself off in terms of how well your mind's eye might work. You can assess your imagery abilities in the next chapter.

Spatial imagery can be particularly affected by damage to the brain's parietal lobes. This part of the brain is important for building mental maps, and Professor Martha Farah of the University of Pennsylvania and her colleagues have studied one man with this type of damage. He could no longer use his mind's eye to locate where he was in space. Because of this, he couldn't describe how to get from his house to the corner store, a trip he had made several times a week for years. He often banged into furniture because he had a very poor mental map of where he was in space. Yet when you compare him with another patient studied by Farah and her colleagues—a man with damage to a different brain area, the temporal lobes—the contrast is telling. While this second man was excellent at finding his way about in space, he couldn't create mental pictures of objects and faces in his mind's eye. If you asked him to describe Abraham Lincoln's face, he simply couldn't do it, and gave a completely wrong picture of this legendary icon. At parties, his wife had to wear some distinctively colored piece of clothing or jewelry so that he recognized her, as he couldn't discriminate her face from others even though his eyesight was fine. In contrast, the first man was fine at recognizing faces and objects, and could create very detailed mental images of them.

These may be extreme cases of the kinds of different abilities in various types of imagery that ordinary people like you and I might show. How can these kinds of differences between people come about?

Picture in your mind's eye the front of the building where you live. Now mentally move around to the back of the building and try to picture that. Choose a room—say a bathroom—whose window is on the back wall. Try to locate exactly where this window is on this wall in relation to the corners of the building as well as in relation to the other windows and doors. Close your eyes as you do this, if you like.

Now do a different visualization exercise. This time visualize the front door of your house or apartment. Try to see in your mind's eye its color, panels, mailbox, and doorknob as a single picture, as vividly as you can make it.

What? Where?

These two exercises depended on two distinct parts of your brain. The first—finding the back window—was based on what is known as the *where* system of your brain. The second exercise—picturing the front door—used the *what* system.

After your eyes have sent their signals to the back of your brain, this information is collated and analyzed in various sections of the occipital lobes at the very back of your head. Then it is fired forward again in two main pathways, a high-road dorsal or *where* route, and a low-road ventral or *what* route. The *where* system arcs up toward the top of your head and forward, curving into and through the parietal lobes on both halves of your brain. In this route, it is *where* the object you are looking at is located that is encrypted in the blizzard of signals rushing forward through your brain. Though this brain system knows *where* something is, it is ignorant of *what* it is.

The *what* information is in the meantime being transmitted down the low road or ventral route. This path goes from the occipital lobes forward into the temporal lobes of your brain—just above your ears on each side of your head. It is in the outside surface of parts of the temporal lobes that the hard drives of memory are located. It's thanks to the information stored here that you can recognize an ostrich, a potato peeler, or a Ferrari. Certain types of dementia, for instance, can lead to a decay of this stored knowledge in memory. When this happens, the patient may lose the knowledge of whole categories of objects—living things, for instance—that are clustered together in these temporal-lobe memory banks.

When you are looking at the front of your home—say you have just rented it and are returning for the first time—you recognize it as yours in part thanks to the low road's ability to compare the front door you see with the one you remember. No doubt the high-road spatial location system also helps you find your new home because it will have encoded information about the spatial layout of the front of the building.

When you are using your mind's eye, you make more use of the low or high road depending on what it is you are trying to visualize. Take a moment to visualize your kitchen, concentrating on where everything is—stove, washing machine, table, window, sink, etc. That's a dorsal route task. Now try to visualize exactly what the kitchen sink looks like. That's a ventral route task, and this slight change in how you visualized your kitchen led to a dramatic change in which parts of your brain were most active.

"Is This a Dagger Which I See Before Me? . . . Come, Let Me Clutch Thee . . ."

Macbeth's guilty hallucination of the blood-dripping dagger was so vivid because the same brain circuits were switched on in the hallucination as would be active if he were actually looking at a dagger dripping with King Duncan's blood.

It seems that we may visualize and imagine by using the very parts of the brain that we would use to see with. This is the conclusion of the work of Stephen Kosslyn of Harvard, who has pioneered research into visual imagery. To help you understand exactly what's going on in your brain when you think in images rather than in words, try this exercise Kosslyn devised.

Take the following letters: a, b, e, h, and r. Closing your eyes for each in turn, decide whether the upper-case version of each letter has any curved lines making it up or not. Take a: A has only straight lines, and so the answer to this is no. The letter b, on the other hand, does have curved lines in its upper-case version: B. Now do this for e, h, and r.

To do this task, you have to leave the world of words and enter the wordless realm of imagery. Your brain has to read *a,* turn that pattern of lines into the sound *a,* and then search in the hard drive of your memory for the upper-case version of the letter. It then has to generate this as a clear mental picture, in order to "see" whether that picture has any curved lines in it or not. This is not something that you know without "thinking" about it, and the only way to "think" about it is by imagery, not through words. Your brain seems to do this type of visualization using the same neural machinery that it would use to scrutinize a real visible B and decide whether or not it had any curved lines in it. In other words, mental "seeing" can closely resemble real seeing in most ways except the activity in the eyes. But this isn't just true of visual pictures—it also applies when you imagine sounds.

Take a moment to imagine the sound of your best friend's voice. Close your eyes and try to re-create the sound of the voice, the words being spoken.

Did you manage to "hear" the voice? If so, it was because your brain activity was very similar to what it would be were your friend actually speaking to you. In particular, a hearing area in the brain called the *auditory association cortex* would be particularly lively. This is located in the temporal lobes of your brain, at the side of your brain just above your ears.

Mental pictures are obviously not straight one-to-one projections from the external world onto a little screen at the back of the head. Nor are they projections of photograph-like images pulled out of the hard drive of memory. You can prove this yourself by trying to imagine a piebald horse sitting in an armchair smoking a pipe and reading a newspaper. Or visualize a jumbo jet flapping its wings like a bird as it takes off. Or imagine you and your chair floating up into the air, and look down at your street from above. Close your eyes while you do this.

It is quite unlikely that you have ever seen such events, but in spite of this you could probably picture these impossible images. So while our mental images are picturelike in some respects, they are put together by complicated codes of firing in the brain's neurons that aren't totally dependent on actual images stored in memory.

Imagine you are looking at your kitchen from the perspective of the kitchen door. Zoom in on the sink. Now move your mind's eye to the fridge. Next the table. Now move your attention to the nearest window. Close your eyes while you do this. Which of these distances are, in reality, longer?

If you really were looking at your kitchen, your eyes would take the longest time to travel the longest distances. And it is exactly the same in the mind's eye. It takes me longer to move my mind's eye from my sink to the window than it does to move it from the sink to the nearby fridge. In other words, the real spatial layout of the world is replicated to some extent on a so-called topographically arranged mental image. And if you imagine mentally walking toward an imagined object, you also find that the mind's eye follows the same rules of perspective as the real eye does.

Imagine an elephant standing in front of you on the other side of the road. Picture yourself walking toward it. You should see the elephant "looming" as you walk toward it, and at some point, depending on the width of your road, the image of the elephant will "overflow" in your mind's eye. Now do the same with a horse, in the same position, on the same road. Mentally walk toward the horse, and again note where on your road the horse "overflows" in the mind's eye.

What you will find—again thanks to the ingenious experiments of Stephen Kosslyn—is that you can walk closer to the mental horse on the imagined road before it fills your mind's eye than you can to the mental elephant. And, of course, this is how it would be were there an elephant standing opposite you. In only a few steps of your approach, the elephant would blot out the houses on the other side of the road, whereas you would have to go right up to the horse before it would obscure your vision.

Had the dripping dagger Macbeth saw in front of his eyes stayed suspended in one place as he moved falteringly toward it, it would have loomed in his vision just as if it were a real dagger. Unfortunately for Macbeth, however, as he strived to clutch it the spectral dagger receded, drawing him toward Duncan's bedroom.

In many ways, imagery is as complicated as language, but because we underuse it, we don't think about it much. Everyone can use some types of imagery better than other types. In the next chapter you can assess how well you can use your mind's eye, and see whether you want to improve it.

4

Do You Think in Words or Pictures?

Paris, 1883: Impressionism is at its height, and Monet is painting light and lilies at Argenteuil. A few miles along the Seine, at Paris's world-famous Salpetrière Hospital, Professor Jean Charcot—a pioneer of hypnosis and a major influence on Freud—opens a letter from a patient, Monsieur X. This successful, intelligent man—fluent in French, German, Spanish, Latin, and ancient Greek—describes how something terrible has happened to his mind. He is now suddenly blind in his mind's eye.

Up until then, M. X could visualize with the vividness of actual seeing. He only had to read something two or three times before it was etched on his mind and he could read it off as if it were chalked on a blackboard. From memory he could make sketches of places he had been as if they were in front of his eyes.

Something has happened. A period of business stress and suddenly, without warning, a bulb has blown in his mental screen. He complains that his previously vivid imagination has gone: "I used to be impressionable, enthusiastic, and I had a vivid imagination. Today

I am calm, cold, and my imagination cannot lead me astray." He mourns the loss of vivid dreams: "Now I only dream in words."

The gift may have been in his genes—his professor brother, orientalist father, and painter sister all have it. Now he can't imagine the faces of his wife and children, he can't let his imagination playfully roam through the wordless realms of image: "If you were to ask me to imagine the towers of the Notre Dame, a grazing sheep, or a ship in distress in the open sea, I would answer that, although I know perfectly well how to distinguish these three very different things and know very well what they are about, they have no meaning for me in terms of internal vision."

M. X was much less affected by grief now. For instance, after one of his relatives died, he could not mentally picture the man's features, his suffering, or the grief of his family. He blames his blunted sorrow on the loss of his mind's eye. Generally, he feels crippled by the loss of his mind's eye: "I now have to say to myself the things I want to remember, whereas in the past I would only have to photograph them by sight."

How Good an Imager Are You?

Brain imaging had not been invented in the 1800s, and so we don't know what happened in M. X's brain that caused him to lose his much-prized capacity for mental pictures. Perhaps he suffered a stroke, or a series of small strokes, without realizing it. This seems likely, as it wasn't just his visualization ability that he had lost—he also had difficulty recognizing faces, and other problems.

This sophisticated, multilingual French businessman is an extreme example of how central imagery can be to our mental life. For M. X, his mind's eye was all-important, not surprisingly, given how exceptional his talents for imagery seem to have been. How much do you rely on visual imagery? How good at it are you? Can you train yourself to be better at it? These are some of the questions for this chapter.

Let's start with your own ability to make mental pictures. In the early 1970s, David Marks of the University of Middlesex in London devised a way of measuring how vividly people can conjure up pictures in their mind's eye. Take a few minutes to do his test.

Vividness of Visual Imagery

In this test think of the numbers you give yourself as ranks—first place, second place, etc. In other words, 1 = good and 5 = poor. The lower your total score, the better your visual imagery. You are rating your visual images in terms of how well they correspond to "mental pictures," how clearly you are "seeing in the mind's eye."

First, get this scoring system clear in your head:

1. The image is perfectly clear and as vivid as normal vision.
2. The image is clear and reasonably vivid.
3. The image is moderately clear and vivid.
4. The image is vague and dim.
5. There is no image at all; you only "know" that you are thinking of an object.

Remember, 1 is best imagery, 5 is worst.

For the first four questions, think of some relative or friend whom you frequently see (but who is not with you at present) and consider carefully the picture that comes before your mind's eye.

scene	rating (1 to 5)
1. The exact contour of face, head, shoulders, and body.	_____
2. Characteristic poses of head, attitudes of body, etc.	_____

3. The precise carriage, length of step, etc. in walking. _____
4. The different colors worn in some familiar clothes. _____

Next, visualize the rising sun. Consider carefully the picture that comes before your mind's eye.

5. The sun rising above the horizon into a hazy sky. _____
6. The sky clears and surrounds the sun with blueness. _____
7. Clouds. A storm blows up, with flashes of lightning. _____
8. A rainbow appears. _____

Think of the front of a store that you often go to. Consider the picture that comes before your mind's eye.

9. The overall appearance of the store from the opposite side of the road. _____
10. A window display including colors, shapes, and details of individual items for sale. _____
11. You are near the entrance. The color, shape, and details of the door. _____
12. You enter the store and go to the counter. The counter assistant serves you. Money changes hands. _____

Finally, think of a country scene that involves trees, mountains, and a lake. Consider and picture that which comes before your mind's eye.

13. The contours of the landscape. _____
14. The color and shape of the trees. _____
15. The color and shape of the lake. _____
16. A strong wind blows on the trees and on the lake, causing waves. _____

Add up your scores for all sixteen items. Those of you with Hollywood-quality mental pictures will have the low scores—best score possible 16—and those of you living in the days of radio will have high scores—worst score possible 80.

How did you fare? Here is a very rough guide for interpreting your scores:

Less than 30:	Above average
30 to 40:	Average
More than 40:	Below average

If you scored roughly in the range 30 to 40, then you are average in this type of mind's eye visualization; if you scored over 40, then you are below average; and the more over 40 you are, the poorer you are at this particular type of mental picturing. If your score was under 30, on the other hand, then you are above average, and the nearer you get to the maximum score of 16, the better your visualization ability is.

I do pretty badly on this questionnaire. When I close my eyes to imagine a friend's face or the sun rising above the horizon, I see a very dim outline with very little color. Most of my ratings are 4: "vague and dim." My total score is in the 60s, which makes me a pretty poor specimen as far as visual imagery is concerned.

Does this mean that my brain's sixth sense—its power of imagery—is dead? I don't think so. In the last chapter I showed you how there are different types of imagery controlled by different parts of the brain. I am not good at what we might call "picture imagery." If, on the other hand, I try to create images in the other senses, I am able to conjure up powerful tastes, touch, and bodily feelings. Take the exercise about touching, then biting into the apple, in Chapter 1, for instance. I can much better imagine the touch of the apple than I can the look of it.

Similarly, I can "visualize" its taste and the sensations of biting into it. And when it comes to imagining my body moving—say leaning out against the wind in a heeled-over sailboat—the imagined feeling of my straining muscles and stretched-out back seems much more vivid than any visual image to do with sailing. Sound fares better than sight, too: I can "hear" in my mind's ear the sound of voices of people I know better than I can imagine their faces.

These anecdotes of mine fit in quite well with what we know about imagery and the brain. In the last chapter I showed you how some aspects of imagery depend on the switching in the same brain areas that would be used to see, feel, or hear in real life. For whatever reason, my brain has difficulty getting the occipital cortex—the vision center of my brain—to make the pattern of brain activity that real scenes produce.

If you did not score highly on David Marks's test, you might still be good at imagining space rather than pictures—in other words, using the "high road" of your brain rather than the "low road" I described in the last chapter. You might, for instance, be good at swiveling images in your mind. Take the letters and numbers on page 39. I find it quite easy to flip them over and rotate them in my mind, and this is true for abstract shapes as well. Yet I can't get a vivid picture in my mind of the letters or shapes that I'm rotating.

People who have good imagery as measured by David Marks's questionnaire seem to be different from less vivid imagers in several other respects. If you had a good imagery score, then, on average and compared with people who had poorer scores on the test, you will: tend to remember more nighttime dreams and have more daydreams; have more of an emotional response to images; and be better at controlling a visual image—for instance, visualizing a car driving slowly past, or a friend walking toward you.

But of course everything here—including the vividness questionnaire—relies on subjective reports. How can we be sure that people aren't just making everything up? Well, objective tests also discriminate. If you are a vivid imager, then on average you will tend to: be more creative (but only if your IQ is above average); be a better proofreader; be better at solving geometric puzzles; have better immediate visual memory for details of a scene; and be more easily hypnotized.

These are all average differences: there are creative people with poor imagery, and many uncreative imagers, but the fact is that the skill of vivid imagery is not just a question of better home entertainment inside your skull—it has very useful practical consequences.

Touch, Taste, Sound, and Scent

Let's take a moment to assess how good you are at creating images in the other senses. Imagery is not just about vision. Being able to create images of bodily sensations is crucially important for learning to fight certain diseases and generally in using the mind to control the functions of your body—more of this in Chapter 8. So if you want to learn to use your brain's imagery powers, you should really take some time to run an audit on your own profile of imagery strengths and weaknesses.

Take a moment to imagine the sound of escaping steam. Try to imagine it in your mind's ear with the vividness of real hearing. Could you "hear" this sound? Now try to imagine the sound of a breeze rustling the leaves of a big tree. It's probably easier if you close your eyes. Try one more sound—imagine the sound of waves lapping onto a gently sloping beach.

Now consider touch. Close your eyes and imagine the touch of someone stroking the back of your neck with a feather. Can you "visualize" this vividly on your mental "touch screen"? Now take a moment to imagine the prick of a pin on your finger. Close your eyes again and imagine the feel of someone's lips on your cheek. Finally, imagine yourself lying on a beach: feel the hot sand on your back and legs, and imagine the sunlight warming your body.

You probably found some of these scenes quite pleasant to imagine, if you gave yourself the time, that is. If you have a spare five minutes at your desk, in a traffic jam, or on a train, you could run through these and similar images: the more you practice imagining in the different senses, the better you will become at it.

As I'll show you later, imagining touch and other skin and bodily sensations is a key element of some types of therapy for stress and illness. If you are under stress, taking some time to visualize beautiful or sensuous sights, sounds, or touch can be a powerful antidote to the buzzing, sparking circuits of stress and tension.

You can test your taste imagery too, by a mental tour around your fridge and kitchen cupboards. Can you imagine vividly the taste and texture of a segment of sweet orange in your mouth? Or the soft, warm yielding of a slice of well-buttered garlic bread? Lay some grains of salt along your mental tongue. Pucker your mental lips with a bold bite into a sour lemon. Whatever your favorite food, take a moment for a mental meal. But beware; it will make you hungry.

Smell is taste's sister. It is also one of the most powerful senses, with a royal road to the memory centers of your brain. That's why you can be dizzied back with such poignant vividness to long-ago events by some evocative scent.

When people rate the vividness of imagery in the various senses, the sense of smell comes out on top, taste comes next, and then hearing. Lowest of all comes visual imagery. But these are just averages: What is your best medium of imagery? It's worth finding out, because you can then build on your strengths and learn to use imagery to control health, boost creativity, and fight stress, as you'll see in later chapters.

There are two other types of imagery. The first is kinesthesis—the senses that tell you about the movement of your body and limbs. Imagine yourself running upstairs. Or swimming. Or stretching and yawning after getting out of bed in the morning. These are examples of kinesthetic imagery. This type of imagery is now a crucial part of the training program of world-class athletes. Its power is quite remarkable, and this has been recognized and incorporated into the training regimes of track athletes and golfers, marksmen and divers—you name the sport, and mental imagery will have been used as a training method. You can guarantee that right now, across the world, thousands of top athletes are perfecting their skills in the kinesthetic mental gym. I'll come back to this later.

You can also imagine more general states of your mind and body. Take a moment, for instance, to picture yourself feeling very drowsy. Imagine the heavy eyelids, the woolly head, and the urge to close your eyes and sleep. Now imagine yourself after a few alcoholic drinks—can you imagine the feelings of mild intoxication? Now picture yourself as razor-sharp and alert—refreshed and on the ball. The average person can imagine such complex mind-body states more vividly than he or she can create visual images. And when you imagine them, you change not only your brain but your body, as powerfully as any drug can.

Words Damage Images

Words and images don't go well together. Once you start naming or talking about things, your language takes over and shoulders aside your underused imagery abilities. For evidence of this, try this exercise. Look at the picture at the beginning of the chapter. You may already have seen that this is an ambiguous figure—in other words, it can be seen as two different animals. However, your brain can only perceive one animal at a time. Take a few moments until you can see both creatures.

Did you get it? You can see this either as a duck or as a rabbit. If visual imagery really is very close to real perception, then you should show the same switching between duck and rabbit in your mind's eye as you do when you actually look at the picture. Try this. Get a clear image of it in your mind, close your eyes, and see if you can get it to switch between duck and rabbit in your mind's eye.

But what happens to your mental image if you bring words into these ambiguous pictures? This is what researchers in Trieste, Italy, asked. They reasoned that, if you could temporarily turn down or switch off the brain's language system, you should release the visual imagery system to produce a better visual image—one that is more like "real" seeing. One way of testing how "seeinglike" an image is, is to see whether the image behaves in the mind's eye as it does in reality. The Italian researchers argued that if the speech system was turned down while people looked at this ambiguous picture, they would see the mental image of this picture "flip" more often. But how do you turn down the language system? One way is known as *articulatory suppression*. This means repeating nonsense syllables (*la-la-la-la,* for instance) over and over. This effectively ties up a key part of the brain's language apparatus—like giving a new toy to children to keep them from getting into mischief.

So what happened to the Italian children and adults who turned down their brains' language systems? Well, of eighteen adults, thirteen who had *la-la*-ed while studying the ambiguous figure saw it switch to the other figure on their mental screen. Only five of another group of eighteen adults, whose language system was left untampered with, saw the image switch.

A group of ten-year-old children showed the same results as the adults—again the ratio was 13:5. But it was quite another story for the six-year-olds. Switching off their language system in this way had no effect on how often they saw the mental image switch. Why? As Robert Graves's poem argued so beautifully, young children have not yet let their mind's eye be dulled by the "cool web of language."

Pictures Painted by Words

Imagery is useful when it comes to words and language too. Many of you have to read for your work; others read for pleasure. Imagery can help—and hinder—how much you take in of what you read and how well you take it in. If you want to get the benefits of imagery in reading at work and for pleasure, you should understand a little about how your brain's language and imagery systems can work together.

Some words can trigger images in all our senses more than others. You can, for instance, visualize the word *rose* more easily than the word *strategy.* The sentence "There are twenty-four hours in the day" doesn't trigger imagery in the way the sentence "An apple is bigger than a grape" does. So though words can short-circuit the work of the mind's eye, they can also—depending on how imageable they themselves are—stimulate visual imagery. Let's say you read a high-imagery sentence such as "A tomato has yellow seeds." This will take you longer, on average, to read and understand than a low-imagery sentence like "Monday occurs about four times per month." But it takes the same time if you listen to these same sentences on an audiotape. How so?

Well, to read the words on this page—any words—you need your mind's eye to some extent. Decoding these black squiggles we call letters into the meanings stored in your brain demands some imagery. If someone read this page aloud to you, on the other hand, your mind's eye wouldn't be needed—it would be your mind's ear that would be called into service. With this "sound card" in your brain being used to decode the words, your mind's eye is much freer to create visual images to the spoken words.

In other words, we listen and read using both the language and imagery circuits of our brains. For very abstract, low-imagery sentences such as "The ticket is only valid for three days," most people probably don't use much visual imagery. But there are other types of abstract statements that do demand imagery.

Take arithmetic, for instance. If I ask you to tell me whether this equation is right or wrong—15 + 19 = 34—many of you will check this by "writing" the sum on your mental screen as if on paper. Others will do it in a nonvisual way, for instance by breaking up 19 into 15 and 4, adding the two 15s to give 30 (15 + 15 is fairly automatic for most people), and then adding the 4, again another pretty automatic piece of arithmetic.

So for most words and sentences, there is more than one way in which your brain can understand them. Often we use a mixture of methods, in different combinations. When you read a story or text, however, your brain's ability to visualize the meanings is slightly sabotaged by the demands the written text itself makes on the mind's eye. Listening to the same story frees up the mind's eye, making it slightly easier to use mental pictures to make sense of the story. I listen to talking books quite a lot, and I find that my impoverished mental imagery is greatly helped, compared to when I read the same books on paper. My anecdotal experience fits with the research findings. You might like to try this yourself to see whether listening to a book gives a more vivid experience than reading it.

Reading a book, of course, will almost always be faster than listening to it. This is because many people tend to skim the page, picking up the gist of the narrative by homing in on key words and sentences—the beginnings and ends of paragraphs for instance. I remember as a student going to a speed-reading course of the type that was in vogue in the 1970s. We were led to believe that you could train your eye and brain to take in whole blocks of text—scores of words at a time—and that reading word by word was primitive and inefficient. It was the course that was a waste of time, however. The proposition that you can take in the full meaning of scores of words in a single glance has been shown to be wrong: yes, you can quickly get the gist of a whole block of text at a single glance, and you can race through a book getting a pretty good idea of what the author is saying. But the faster you go, the more you miss. So if friends boast that they can read thick novels in an afternoon, test them on what they remember of the detail. You probably can glean the plot by racing through a novel. But you'll miss much of the subtleties of the language, the scenes, and the narrative.

We read different books for different reasons. I couldn't possibly read all the books and papers I have to read in an average day without some very rapid, strategic skim-reading. But some books and papers I have to read very closely, and if I am reading for pleasure, I see little point in romping across the text gleaning only the gist. One reason that I prefer listening to books to reading them—apart from the fact that my mind's eye is freer to generate its feeble images—is that I take in every word and can enjoy the elegance and style of the author.

Whether your brain visualizes words or not doesn't entirely depend on the words themselves, however. You can use your mind's eye to a greater or lesser extent when reading or listening to words, and your brain activity will be different depending on how much you decide to visualize. Volunteers in one study were asked to read words normally, and then to read them while making a mental image of the words' meanings. The volunteers read the word *cat,* for instance, and then they read it while creating a visual image of the cat. Researchers compared the electrical activity in the brain when the mind's eye was "switched on" with that when it was not. When they were visualizing, there was much more brain activity in the occipital lobes of the brain—the vision centers—showing that one key part of the brain's visual-imagery apparatus could indeed be deliberately switched on.

This is good news if you want to boost your understanding and memory of what you read. Some of us naturally create visual images while reading or listening, but many of us don't tend to do this much, even though we could if we tried. You might not want to clutter your reading with visual imagery, but if you do, the scientific research shows quite clearly that you can tune up your mind's eye to produce a more vivid response to the written and spoken word in your brain.

Language, of course, is not just about the meanings it conveys directly through the words used. Particularly in poetry, spoken language has its own music, and just as we can visualize the face of a friend or relative, so too can we use the mind's ear to imagine the sound of that person's voice.

Without actually speaking, say the words of a poem or song—a national anthem if you like—in your head. Now imagine these same words being spoken to you in another person's voice—say that of a close friend or relative.

These two different mental exercises produced quite different patterns of activity in your brain. In the first case, where you mentally spoke the words yourself, the main brain areas activated were in the speech-production centers in the lower part of the left frontal lobe, down on the surface of your brain near your left eye. When you imagined your friend speaking these words, however, your brain's speech-reception areas sprang into life, farther back in the left half of your brain, near your left ear. In other words, your brain was responding as if you were actually hearing that person's voice.

Imagining voices like this is something that most of us do a lot of the time. But if the brain's activity in response to imagined voices is much the same as its activity in response to real voices, how on earth do we distinguish between reality and imagination? Well, in some abnormal mental states, this is not so easy. In states of grief, for instance, the grieving person fairly often hallucinates the lost person as if he or she were really there. Certain drugs—LSD is an example—can also produce vivid visual and auditory hallucinations. And many schizophrenics are tortured with voices that can seem utterly real to them.

Normally we can tell the difference between voices inside and outside of our head, however. This happens because we monitor the source of the words. So when we daydream and imagine sounds and voices, we also monitor automatically whether we have triggered them or whether they have come from somewhere outside. If the brain registers that the sound or voice was triggered internally, a message is sent to diminish somewhat the level of neural activity in the speech-reception area of the brain. When we hear a voice that we have not generated—i.e., a real one from outside of our head—the activity in the speech-reception area is not turned down. This is partly how we tell the difference between the real and imagined voice. With some types of schizophrenia, this "turning down" doesn't happen, and so sufferers find it hard to know whether or not what they are hearing is an inner voice or a real voice.

Not only can words project pictures on the mind's eye, then—they can also fill the mind's ear with sounds. Some poets cater more for the mind's ear, others for the eye. Others write in highly abstract language that does not easily generate visual or other images. Poetry changes the brain in different ways, but most poets trade heavily on the currency of imagery. What happens, though, to the imagery of people who have been blind from birth: can they see in the mind's eye?

And the Blind Shall See

What *does* happen when you ask a man who has been blind from birth to visualize some scene? Surely he will give a despairing shake of the head at your insensitivity? No, he won't, because congenitally blind people can create powerful mental images that have a lot in common with visual images. Take dreaming, for instance. If you have never, ever seen anything in your life, have no visual memories, no experience of color, of visual movement, of visual shape, surely your dreaming must be confined to sounds, touch, smell, taste, and the feelings associated with your body's sensations?

What, then, do you make of this? A blind person is awakened at night in the middle of the dream phase of sleep. When the researchers ask her to say what she was dreaming about, she describes a room in a cancer clinic, in which there is a big machine with buttons on it, something like an ATM. On this machine there is a screen on which staff can see how the various patients were doing. It is a large room, oblong in shape.

When they asked her whether she could "see" this room and its screen, she said she couldn't, but she knew that *others* could see the screen. Nor was she using touch imagery of the ATM-like machine by imagining her fingers touching the buttons. Rather, she knew where the buttons were in relation to the rest of the machine without touching them. This wasn't a dream of colors and visual vividness, but it was a coherent description of a layout of a room and its equipment, including relative sizes and locations. This was a dream of spatial images—of shapes, positions, and sizes—but stripped of their visual qualities of color and shading.

In the last chapter, I showed you how this type of spatial imagery is located in different brain areas from classic picture-like visual imagery. While people blind from birth have no experience of the subjective experience of seeing, they have plenty of experience of space. They move around furniture in their houses, navigate themselves around streets, reach out for pens and books at their desk, and reach to switch off the radio on their bedside table. They use mental maps that have the outlines and spatial layout but not the visual detail of all the features.

So to a certain extent, the blind *can* "see" in the mind's eye, because they can create a colorless "picture" of objects and spaces through the assembly of abstract spatial copies that preserve the relationships between the elements of the imagined scenes. Maybe that's what I—a nonvivid imager—am doing when I can "see" someone's face without having a vivid "picture" of it in my mind. Perhaps what I am "seeing" are certain abstract spatial properties of the face—line of hair, curve of mouth, shape of nose, slant of eyes—that are not specifically visual.

Whether that is the case or not, other studies show clearly that congenitally blind people visualize in very, very similar ways to sighted people. Take the rotated letter experiment that you did in the last chapter, for example—the one where you had to say whether the letters and numbers were mirror-reversed or not. You do this by mentally rotating them to an upright position in your mind's eye. Blind people show the same pattern of results as nonblind people. When they touch raised letters of this type, they too have to rotate them on their mental screen, and they too take longer to decide whether it is mirror-reversed or not the bigger the angle they have to rotate it through. Similarly, if you ask them to explore by touch a set of objects on a tray, they can memorize this layout as an abstract spatial image. This image has a huge amount in common with the visual image that you and I would create of the same board.

One classic experiment on visual imagery involves imagining an object in the context of a much bigger or much smaller object. You then ask questions about the first object. Imagine, for instance, a cat sitting beside a car. Now imagine the same cat sitting beside a paper clip. You are then asked a question about the cat in each of these conditions—for example, "Do the cat's ears stick out above its head?"

It is a well-established finding that it takes longer to answer this question when the cat is dwarfed by the car on the mental screen than when the cat looms over the paper clip. This is because of the time it takes to "zoom in" on the cat on the mental screen with the car on it, to check the ears on your memory-stored snapshot of a cat. Try the same exercise by imagining—again in the context of either a car or a paper clip—a rabbit. Where are the rabbit's ears in relation to the line of the head?

Congenitally blind people showed the same pattern of results—longer reaction times to "zoom in" on the object beside the car—as sighted people. Again, this shows an uncanny overlap between the mental "pictures" built up in the minds of people who have never had the experience of seeing anything and those built up in the minds of the sighted.

This finding might help us understand a rare but perplexing problem that can arise when the visual centers of the brain are destroyed in an accident or disease. Sometimes when this happens, the person denies that he or she is blind, even though on objective tests they demonstrably are. Even though they may find themselves falling over furniture, not seeing their meals on the table, and ignoring the visitors at the bedside, they still maintain the conviction that they can see.

This phenomenon of *anosognosia* (not recognizing one's illness) for blindness is not fully understood, and there are many complex factors that play a part in it. One set of researchers in Germany studied a woman who had lost most of her occipital cortex, the vision area of the brain, but who insisted she was not blind. Like congenitally blind people, she had a reasonably good capacity for visual imagery. The researchers studying this woman suggested that she might be mistaking her mental images for real perceptions and that this was one reason why she clung tenaciously to the conviction that she could see.

They also thought that she might be experiencing *synesthesia*—namely, experiencing sensation in one sensory modality via another. It is possible that touching things, for instance, triggered visual images, probably in the remnants of her occipital cortex that had not been damaged. From her perspective, "seeing is believing," and she found it hard to accept that she was blind when she "saw" things as she touched them.

Significant support for this came when researchers in London studied people who had lost their sight because of damage or disease to their eyes rather than to their brains. Using PET brain-scanning methods, which highlight which part of the brain is active during different mental and physical tasks, these people were studied as they used their fingers to read Braille. What the London researchers found was that the primary visual areas of the brain became active, in spite of the fact that the subjects were only "feeling" and not seeing. It seems that the touch of the Braille letters triggered visual images as the subjects would have seen them before they went blind. This may have been happening to the person who denied her blindness, giving her additional ammunition in her belief that she could see.

Therefore the blind can, in a certain sense of the word, "see" in the mind's eye. Or maybe we should put it another way: perhaps a lot of what sighted people "see" is really this kind of more abstract spatial representation, not tied to any one sensory modality.

Can I Make Myself a Better Visualizer?

Over the next few chapters you will see that imagery has a number of very practical applications in everyday life—for instance, in memory, creativity, stress reduction, and health. So for some of us at least, the question of whether you can become a better visualizer is worth asking.

First, what about people who have to visualize in order to survive. Architects and engineers, of course, have to be able to manipulate space and shapes on their mental screens, and they won't be very good at their job if they aren't able to do this. Artists such as painters and sculptors will—at least in the classical as opposed to the conceptual forms of the art—have to be able to visualize and manipulate images in the mind's eye.

But there is one group of people who may have to learn to create mental images highly effectively even for basic communication with other people in their lives. People who are born deaf in most cases learn sign language. American Sign Language is one such communication system, with the distinctive feature that meaning is partly conveyed by the position in space where certain signs are made by the hands. In other words, imagined space is absolutely central to deaf people being able to communicate: They have to be able to lay out the people, objects, and places they are talking about in their mind's eye in order to convey what they want to say about them to the people with whom they are communicating. What's more, some movements have to be read from the perspective of the "listener" and others from the perspective of the "talker." So, for instance, suppose a deaf woman wants to describe a particular room. First she gives the sign for "I enter" to convey that the scene she is describing should be seen from her own point of view. Then, to convey that there is a table on the left of the room from her perspective, she makes the hand sign for a table in the air to the left. If there is a chair on the right, then she will make the sign for a chair in the air on the right. This is just a very simple example. Much more complex actions and descriptions are set out using space like a mental blackboard. And, of course, the space is entirely imaginary—both the speaker and the listener using ASL have to create the same mental images in order to communicate.

The point is that deaf people who have learned ASL depend on good mental-imagery ability for the most fundamental of human needs—communication with other people. And as such, they have hundreds of thousands of hours of training in using their minds' eyes with pinpoint accuracy and high levels of complexity. If mental imagery can be trained, surely these people must be masters of the art?

Well, at the very least they are a lot better at—as you might expect—the *spatial* aspects of imagery. Let's take mental rotation, for instance. Let's say you are "listening" to the woman "speaking" to you using ASL, describing the room with the table and chair in it. If she is describing the room from her point of view, to understand her description you have to mentally turn yourself around to get the scene from her perspective. As you know from the last chapter, this sort of mental rotation takes time and mental effort, as your brain has to recalibrate all the points it is representing according to a new map. But those who are fluent ASL users, who are used to listening to others speak ASL, are able to rotate themselves mentally to a new perspective with a fraction of the effort of people whose brains have not been trained in this particular imagery method. Some nondeaf people—mainly family members of someone who is congenitally deaf, or professional people who work with the deaf—are very practiced ASL users. Not surprisingly, given the extensive training that their mind's eye has had, they also show the same types of highly efficient and accurate mental imagery as the deaf ASL users.

American Sign Language doesn't just give practice in mental rotation, however. If you wanted to communicate that someone was going to lie down, then you would signal both the direction and the final height of the movement in the mutually imagined space between listener and speaker. Highly complex scenes and concepts are conveyed in this way, and so you should expect—if imagery can indeed be trained—that ASL users would be much better than ordinary people at generating pictures in the mind's eye. And so indeed it is, for many but not all aspects of mental imagery. It seems in particular that spatial imagery, of the type that blind people seem to be capable of, is what ASL users excel at. They generate mental images faster and have a better short-term memory for spatial patterns.

It seems, then, that you *can* improve how well you use your mind's eye. More evidence for this comes from a study of London taxi drivers. They—possibly uniquely among taxi drivers in the world's big cities—have to pass a stringent test called the Knowledge. They have to learn by rote the location and layout of a vast amount of the streets in this sprawling city. To learn this, they spend a minimum of two years touring the city on mopeds, with a clipboard on the handlebars. They also have to learn the layout so that they can take the geographically shortest route between any two points—not as easy as it sounds with a meandering River Thames and odd-angled streets. If they manage to pass this stringent test of visual and spatial memory, most of them manage to transport you about London without the fumbling with street maps and frequent stops that characterize the behavior of taxi drivers in many other places. In other words, the cab drivers of London are constantly exercising and stimulating their brain's capacity to remember, recall, and use mental maps of the layout of this great city.

Researchers in London have studied the brains of more and less experienced taxi drivers and found that one key brain area for this type of nonlinguistic, spatial memory—part of the hippocampus—was more enlarged in the experienced drivers than in their less experienced colleagues. So not only does practicing using the mind's eye improve your imagery abilities—as in the case of the sign-language users—it can physically expand certain parts of the brain underpinning the type of imagery practiced.

Of course, mental imagery is not a single entity, but rather a whole sequence of different mental operations carried out by different parts of the brain. And individuals differ in how well the various parts of the machinery of the mind's eye function. Indeed, there is good evidence that if you want to sharpen your mind's eye, you will have to oil up the machinery of several different parts of the imagery process. Practice does improve visualization, but you have to make sure to practice the separate elements of the complicated business of imagery. If you feel so inclined, you can do some exercises that might help improve the vividness of your mental pictures. These exercises focus on just one component of visualization—the ability to keep a mental picture alive. Stephen Kosslyn calls this the *regenerate* mechanism of the mind's eye.

You can choose any class of objects for this exercise, but just for illustration, let's stick to animals such as the horse, chicken, dog, duck, elephant, fox, giraffe, gorilla, pig, skunk, and squirrel.

Now imagine a grid of 16 squares set out in a 4-by-4 matrix. You are to "project" all the animals onto this grid when you visualize them. Also, you have to learn to pinpoint squares on the grid, which is marked A, B, C, and D along the top, and 1, 2, 3, and 4 down the side. The top left square is A1, the bottom right is D4. Close your eyes and imagine this grid, and practice labeling the squares.

Pick one of the animals, imagine it on the grid in your mind's eye, and then pick an area of the body—the end of its tail, for instance—and pinpoint in approximately which square on the grid it is located. Take an elephant facing left, for instance. The end of its tail will be somewhere down around square D3, whereas for a chicken, it will be up nearer D1. You can practice this for various body parts—nose, tip of the ears, bottom of the neck.

The better you scored in the Vividness of Visual Imagery questionnaire on pages 53–54, the better you are likely to be at this type of task, even before you do some mental imagery training. If you take time to practice this exercise, you'll get better at other types of mental imagery that depend on your brain's ability to work with visual images.

If you feel inclined, practice the animal task when you have an odd idle moment, waiting in a traffic jam, for instance, or sitting on a plane. People who practice this type of task show improvements in certain aspects of mental imagery, though not, for instance, in the manipulation, movement, and *control* of mental images such as by mental rotation. This depends on different components of the machinery of the mind's eye and you have to practice these separately.

In my book *Mind Sculpture,* I showed that the human brain can be physically shaped by experience, and that our mental capacities can—within limits—be improved by practice. Imagery seems to be no exception to this. What are the advantages of improved visualization? Well, let's turn now to a key skill that needs good imagery—creative thinking.

5

Better Imagery—More Creativity

She lies on her side, one hand resting lightly on her swollen belly. The ligaments of her body are visibly slackened, ready for the brutal stretching of birth: the loosened tissues allow her hips to curl yielding and folded around the filled womb. Her eyes are only half-focused—apparently on some place or thought beyond the deep gorge of child-birth. Her expression is not quite sad, not quite anxious, not quite the joy of impending motherhood. No, it is more one of a sort of fulfilled resignation, an awareness of many sorts of pain and tinged with the ineffable aura of belonging. Such living detail is chiseled out of cold, hard stone, right down to the faintest wrinkles etched around the youthful eyes.

An egg-shaped block of stone—four sweeping arcs polished and un-creased by the slightest detail. That the sheer rounded simplicity of this block should convey "woman" is miraculous. But it is not only woman—it is archetypal, canonical woman curled protectively around her womb. Strangely, it is even more woman than the re-

clining pregnant woman of the neighboring sculpture. Here is an essence caught precisely by the sidestepping of detail.

Two schools, two artists of the human form, you may think—the first a Michelangelo, the second a Henry Moore? No. The same artist carved both—the first when her brain had not been damaged by the disease lupus, the second after it had. But this was no change in artistic style kindled by the emotional challenge of illness. No, this woman had lost her mind's eye. She had lost it because of damage to the brain circuits where a vivid visual image of her pregnant model was once held. Held in a shimmering stillness, in exquisite detail, so that she could re-create it beautifully with her sculptor's hands.

But strangely, this loss of sight in the mind's eye seems, if anything, to have enhanced her creativity rather than diminished it. Fine though her early work was, the raw, essential simplicity of her post-lupus art cuts through the detail and captures the magnificent essence of her subject. But how? How can one re-create the visual form if the vessel transporting it from life to clay is shattered?

Creativity means forging unthought-of links between ideas, images, and senses. This artist learned to "see" with her hands. It was as if she had come to use a new vessel to carry the form from life to clay—the felt shape of the model, cupped in mental hands. And as the form was transmuted from seen to felt shape, so it was stripped of the clutter of the eye and an essence captured in raw sweeps of stone.

The Leg-Irons of Mind-Habit

Nobody can get through the average day without habit. Most of what we do, think, and say happens automatically, mostly outside of awareness. If we had consciously to ponder and decide about these responses and reactions, our brains would collapse from information overload in seconds. Habits of speech, of thought, of emotion, and of perception are our savior—but they can also be a curse.

When we are young children, the habit of language—the "cool web"—hasn't yet pushed aside our native ability to create vivid mental images of the world. And it seems that such mental imagery has been central to some of the great acts of human creativity, both in art and in science. But at school we learn more and more to rely on habits of language to respond to the world, and less and less on imagery—though this was not the case at the school Albert Einstein attended, as you'll see in a moment. However, precisely because imagery tends to be underused, it tends to be less habitual, less automatic—and hence, potentially at least, more flexible. In short, the fact that we underuse visual imagery means that it may be a potent route to creativity when we do use it.

Mental imagery is not—as the last four chapters have shown—a single mental process. The sculptor whose art changed after her brain was damaged clearly had not lost in its entirety her childhood ability to create and copy vivid mental images. Before her illness, she captured the finest detail of these images. After the disease, her brain shook off the detail and instead imaged the fundamental shapes underlying the forms she was sculpting.

The sculptor had learned to rely on parts of the brain that had not been damaged by the disease. These were very likely the same circuits that congenitally blind people use to "see" in their mind's eye—mental pictures composed of abstract form and shape, and not of visual detail. And by shaking off the detail of vision, she penetrated to the deep currents of form that are so often obscured by the superficialities of detail.

She could gain access to these currents by crossing the boundaries of the senses. For instance, the researchers who studied her—Barbara Wilson and her colleagues in Cambridge, England—gave her tests of visual imagery, asking questions like "Does the Star of David have five points?" Now, it is generally assumed that most people answer this by making a visual image of the star and counting the points. But our sculptor couldn't do this. What she did instead was to "draw" the star on the table with her finger. When asked not to use her finger, she tried to draw it on the floor with her foot. And when she wasn't allowed to do that, she persisted by trying to trace the star shape with movements of her head. This transfiguration in her mind from vision to touch and movement very likely formed part of the creative revolution in her work. The loss of her mind's eye forced her to use other mind-media, such as touch and movement, and the result was a tremendous surge of development in her artistic talent.

Many of the great artists of this century have tried to do something similar. Only they have had to try to suppress the jaded habits of seeing in order to unleash the brain's capacity for abstract representation of shape and form. Think of Matisse and his figures dancing in a circle—a few ellipses that convey the *essence* of dance in a way a realistic depiction could never do. Or Picasso, with the disjointed lips, teeth, tongues, eyes, and noses of the dying citizens of Guernica conveying more powerfully the *essence* of these human organs and what they express than any faithfully detailed rendering could do.

It is not just in language that the leg-irons of mind-habit can cripple our ability to make the creative leap. Any mental faculty that becomes automatic—including the very act of seeing—can become ossified and impervious to freshness and originality. James Joyce revolutionized literature by overthrowing certain tired habits of language, to the extent of creating a whole new vocabulary in his last novel, *Finnegans Wake*. Where there are new words, there can be no habit attached to them, and so the author can try to penetrate the layers of mind-habit to provoke fresh and untried mental responses in the reader.

There can only be one Joyce in a century—it requires a rare genius indeed to tamper with the "cool web of language" in this way. But you and I can learn to be more creative than we normally are, and it seems that imagery—perhaps because it is not jaded by habitual use—is a pretty useful mental tool for this. Still, to benefit from this tool, many of us have to learn—no, *re*learn—to use it.

Creativity Released by Dementia

Perhaps even the darkest of clouds can have a silver lining. Even within the bleak storm clouds of dementia there may be a nugget of comfort for some people. As was the case for the sculptor described at the beginning of this chapter, damage to the brain may do what great artists throughout the world constantly struggle to do—shake off the shackles of perception and habitual thought to create some quite new vision of the world.

Professor Bruce Miller and his colleagues at the University of California at San Francisco studied every patient seen at a particular clinic in Los Angeles who had been diagnosed with a particular type of dementia known as *frontotemporal dementia*. In this disorder, the disease particularly affects the frontal and temporal lobes of the brain.

All of these unfortunate individuals, struck by illness at ages ranging from forty-four to seventy, showed a loss of mental capacity including worsening memory, trouble with speech, and in some cases change in personality. Of the sixty-nine people surveyed, however, twelve showed either a preservation of creative ability—artistic or musical—or, astonishingly, developed quite new talents in these areas.

Take one of these people. When assessed at age seventy-eight, he had suffered from the illness for almost a decade. He was stooped, quiet, and remote. He found it hard to find words and engaged in painstaking circumlocutions to make himself understood. Though he had not had much formal education, he had been a gifted linguist, learning Chinese, Italian, and several Russian languages with such a gift for accent that he was often mistaken for a native speaker. Though he had had little musical training, suddenly at the age of sixty-eight he began to compose classical music. He described his mind as being "taken over" while he was composing and he "sensed" various tonal intervals that he put to music. Some of his compositions were performed publicly, and his composing continued even after his linguistic abilities had begun to desert him as the disease steadily ate away at the language centers in the left temporal lobe of his brain.

Of these twelve people, no fewer than seven had actually developed *new* skills and creativity as their dementia developed—the other five kept up at a high level skills of this type that they had had before the illness struck. Of the seven in whom new creative talents blossomed as the dementia emerged, five developed visual artistic talents and interests, while two expanded their musical creativity. One among these individuals, for instance, began to experience uncannily and painfully vivid visual images, while another two seemed to develop a remarkable acuity of perception that would allow them, for example, to pick out in a flash a particular coin from a large heap of small change.

What distinguished these people from others with the same illness but no signs of new talents? Well, three-quarters of the talented group had dementia mainly affecting the left hemisphere of the brain, whereas only one-quarter of the no-talent group had mainly left-brain disease. In other words, it was those whose brain language systems had been most damaged that seemed to show these preserved or new talents.

Bruce Miller and his colleagues suggest that centers in the frontal and temporal parts of the left brain may, under normal circumstances, inhibit other parts of the brain that are important in visual and musical creativity, particularly those in the right half of the brain. When these left-brain inhibiting circuits are damaged by the disease, then the metaphorical brake is released from the sister brain circuits underlying visual and musical imagery.

This all raises one obvious but crucially important question. If damage to one part of the brain can—by taking the brake off other parts that it normally keeps in check—release new creative talent, are there other less drastic ways in which we can release this talent? In other words, can we train our own brains to release these brakes and improve creativity? Before going on to answer this, let's take a look at how visual imagery may have contributed to groundbreaking feats of genius.

Visions of Genius

Albert Einstein was one of a number of scientists from the German-language tradition who revolutionized scientific thinking in this century, in surely one of the great communal acts of creation of the human mind. Was it a mere coincidence that the prevailing philosophy of German science and education at the turn of the century emphasized the importance of visual imagery in science and engineering?

Einstein attended the Canton school in Aarau, Switzerland,

which was set up by followers of the Swiss education pioneer Johann Pestalozzi. In his book *Insights of Genius,* the historian of science Arthur Miller of University College London pinpoints this influence as central to Einstein's intellectual development. To Pestalozzi, imagery was the foundation of all knowledge, and visual thinking a fundamental and powerful feature of the mind. According to this view, artists and scientists re-create the world by making images, and indeed Einstein himself realized that his own mode for creative thinking was visual imagery. In 1895, at the age of sixteen, while still at school in Switzerland, Einstein used visual imagery to create one of the fundamental thought experiments of turn-of-the-century physics, an experiment in imagery that led to the conclusion that the speed of light is always constant, irrespective of any relative movement between the observer and the source of light.

The sixteen-year-old's experiment visualized a cart in which the observer sits, chasing a point of light on a light wave. The assumption of Newtonian physics was that as the cart reached the speed of the point of light, it would be as if two trains were traveling at the same high speed: to passengers on the two trains looking at each other through the windows, it would seem that the two trains were not moving relative to each other. If this were true of the point of light on the light wave, however, Einstein visualized that the observer on the cart would just see the point of light bobbing vertically up and down like a cork on the wave, and not moving forward at all.

On the basis of the intuition born of this visual image—a point on a light wave bobbing vertically up and down and not moving forward relative to a scientist careering along on a cart beside it at the same speed—Einstein immediately rejected the Newtonian option. Visual imagery made him conclude that the speed of light is constant—eternally *c*—irrespective of how fast the light's watcher is traveling. And it turns out he was right.

Of course, it is only a hypothesis that the visual-imagery-dominated ethos of his school and of the German scientific world of that time was responsible for this fundamental insight. Einstein was a genius, and may well have come up with that insight even without such an education. But the fact is that Einstein himself declared visual imagery to be fundamental to his scientific thinking. Einstein went so far as to say, "Words or language . . . do not seem to play any role in my mechanism of thought." Rather, he said, "My elements of thought are . . . images. . . ."

Perhaps the most famous example of visual imagery underlying a great act of scientific creative thinking is in the scientist Kekulé's dream that led to the discovery of the structure of the benzene ring. Dozing by the fire, Kekulé saw atoms gamboling in his mind's eye. Then they joined into long strings, twisting like snakes. Suddenly, he noticed one of the snakes seize hold of its own tail. In a flash he awoke: this visual image had unlocked one of the major scientific puzzles of the day—that the benzene molecule had a ring structure.

Letting Off the Brake: Can You Improve Your Creativity?

Imagery isn't the key to all creativity, of course. Mental pictures can help shake off the leg-irons of mind-habit, but just as language has its limitations, so has imagery. In physics and chemistry, for instance, many of the visual images that helped crack the code of matter have steadily been replaced by mathematical formulations that are simply not visualizable. In our more mundane worlds, however, far behind the frontiers of science, it is clear that the mind's eye and its compatriots in the other senses offer us a very powerful companion—and, at times, antidote—

to language when it comes to solving problems and thinking creatively. But can we really use the mind's eye to create new combinations of ideas?

Imagine the letter D. Now picture in your mind the letter J. In your mind's eye, turn the D 90 degrees counterclockwise—onto its back. Now place the capital letter J at the bottom of the rotated D. Can you recognize anything that emerges? Study this in your mind's eye until you get it. If you don't get it, I'll give you the answer shortly.

Try another one. Picture the letter B. Turn it 90 degrees on its side as you did for the D in the last example. Add a triangle to the bottom of the rotated B, the same width and pointing down. Remove the horizontal line. What do you see?

Now this isn't exactly the apogee of creativity, but what this experiment devised by Ronald Finke, Steven Pinker, and Martha Farah showed is that mental pictures in the mind's eye are not just the static end product of some more fundamental thought process. Rather, this and many other experiments show that it is possible to use mental images to "discover" new objects and ideas.

You may well have "seen" that the D and the J made an umbrella shape when you combined their visual images, and that the B and the triangle made a heart shape. This means that these images were not just unchangeable end products of some other process. No, if that were the case, then you would not be able to "discover" the new images from the combination of these separately imaged letters. In other words, mental imagery is—to some extent at least—picturelike in the way it operates. But seeing the umbrella made out of the D and J is not particularly creative, particularly given that I gave you precise instructions as to how to move the letters in your mind's eye. Can we come up with really novel or creative combinations by "doodling" on the mental screen? Ronald Finke tackled this in a more stringent probe into creativity and imagery, as he and his team studied how people came up with new abstract inventions.

Take some simple shapes—circle, square, triangle, and rectangle. Then some letters—D, C, J, P, V, X, L—and throw in the number 8. Pick any three of these at random—you can take two the same if you want. Now, in your mind's eye, conjure with these to see what you can create. For example, given a circle, a D, and a number 8, you might come up with a face, the D on the back being the mouth, and the 8 on its side being the eyes.

That would not, however, be a particularly original combination. What might you create for the letters D and T and the number 8? Try that. You don't have to keep these letters or numbers at any particular size relative to one another. Close your eyes and toy with that combination of objects for a minute or two.

Lessons from the Creative

Even these artificial examples with shapes and numbers show that the mind's eye is a potentially rich hunting ground for creativity. Take the picture at the beginning of this chapter; you can see one person's creative response to this challenge of finding an original combination of the D, T, and 8—a croquet set! Not that this guarantees originality—visual clichés exist too. But perhaps we are less constrained by cliché when we use the mind's eye than when we use words: clichés, after all, become clichés through thoughtless overuse. But many—maybe most—of us underuse the mind's eye and so perhaps there is more opportunity for freshness of thought when we do cultivate it.

When we leave the rather artificial world of the laboratory

and study what distinguishes the more creative artist from the less creative one, we can look for some hints about how to find that elusive freshness and originality of thought. Take this study, for instance. A group of art students was extensively studied and followed up over several years. The question was—could the researchers spot early signs of creative talent in the students that would predict creative success years later? As part of the assessment, researchers carefully watched students as they made their preparations to paint a still life from among a collection of objects on a table.

After they had finished, experts rated their work for creativity. Were there any differences between the highly and the less creative artists in how they set about preparing their work? Indeed there were. For a start, the highly creative artists "played" with many more objects before they began to paint. And "play" they certainly did: unlike the less creative painters, who tended to look at one or two objects and then begin to paint, the highly creative ones picked up objects, stroked them, shook them, listened to them, tossed them in the air, smelled them—even bit into them!

You could also say that the creative still-life painters were trying to escape the obvious and the mundane in the objects before them. They seemed to be trying to struggle free from the cool, easy web of everyday associations so that they could get closer to the word-free sensations and *thereness* of the ordinary articles on the table. It was as if they were trying to smuggle these objects past jaded officials policing the gates of perception. Once through the checkpoint they could take these unfettered images and probe, tease, and change them in the utterly unconstrained infinity of ways that is only possible in the mind's eye—as well as, of course, in the mind's ear, nose, tongue, and hand.

Words Can Corrode the Creative Spark

At some time you will have struggled with a problem or for an idea, and then given up, frustrated. Have you ever found that after giving up the fight and ignoring the problem for a while, the answer or the idea pops into your head out of the blue? This is a pretty common experience. It happens because most of our brainwork takes place outside of the narrow realm of our conscious awareness. Indeed, conscious awareness—and the language apparatus that it is so closely tied to—can be an obstacle to problem solving and idea generation.

Imagine these objects: a circle, a half-sphere, a cube, a cone, a flexible tube, and a ring. Now I am going to name three of these objects and I'd like you to create in your mind's eye something made from these three parts that might be useful. You can bend or deform only the tube, and you can put parts inside each other; they can be hollow and they can be made from any material.

With the following three shapes, I'd like you to make an article of furniture: cone, half-sphere, ring. Remember, you are not to tie yourself down to any particular type of furniture—simply try to create a useful, furniture-like object from these three shapes. Take a minute or two to do this.

Now, try to imagine a weapon made from the following: circle, cube, and tube. Close your eyes and work on these three shapes, again not committing yourself to any particular type of weapon—just try to combine these objects in some way and then play in your mind's eye with how the combination might be used as a weapon. If it doesn't work, try some other combination.

To give an example, with the furniture, I imagined a half-sphere with the cone stuck onto its flat side, and a moveable ring sticking out from the same flat surface. This was an adjustable reclining chair: you sat on the flat surface, leaned your back against the cone, and your feet rested on the ring, which could prop the chair back at any angle desired. With the bottom of the seat being curved, you could also rock back and forward, as the connection between the ring and the flat surface could be made springy and yielding.

My article of furniture may not have been particularly creative, I confess, but the research on such fascinating exercises developed by Ronald Finke and his colleagues at Texas A&M University has shown clearly how creativity and inventiveness can be boosted by quite simple methods. The most creative objects emerged when Finke's volunteers had no idea in advance what shapes they were to use or what class of object they were to produce. Their brains were at their creative best when they couldn't narrow down thoughts and images in advance by preconceptions or conventional associations. By springing the shapes on the volunteers at the last minute, the researchers forced them first to create abstract, not yet meaningful objects. Only once this totally meaningless abstract object was composed in the mind's eye did they try to imagine how it could be used as a piece of furniture, a weapon, or whatever category of object had then been sprung on them.

Inventing objects like this may not be of immediate practical use to you. But using the principle of inventiveness discovered by Finke and his colleagues might. This principle is to try to create new ideas without being tied down to specifics. This is because the familiar lurks in these specifics, and in the familiar skulk the clichés and mind-habits that shackle you creatively. Imagery can be one way of playing with concepts in a way that keeps mind-dulling familiarity at bay. Language, on the other hand, can often be the vehicle for drawing you into well-trodden paths of thinking that are very difficult to shake off. Words also dull your ability to solve certain types of problems—

so-called insight puzzles. These are tasks where the answer typically comes to you "in a flash"—an unbidden bolt of intuition from your unconscious mind. Here are some examples.

A prisoner was trying to escape from his cell. He found a rope that was only half long enough to reach the ground. So he divided the rope in half, tied the two parts together, and escaped. How?

Or try this one. A man offers an antique dealer a beautiful bronze coin, with the emperor's head on one side and the date 532 B.C. stamped on the other. The dealer examined the coin, but instead of buying it, he called the police. Why?

Finally, try this. Two ropes are hanging from a ceiling, but too far apart for you to hold both at the same time. Your task is to join them. You only have a screwdriver, a pair of pliers, and some thumbtacks. How do you join them?

Words are a handicap in solving these types of problems: it doesn't help to set about them with logical statements and reasoning. Quite different brain processes are needed here, and research has shown that you tend to interfere with these processes if you use language to try to talk yourself logically through the problems. Solving them requires a kind of lateral thinking style that doesn't fit with words and language. The more you tend to use imagery, the better you will be at coming up with a solution to this type of problem. When you are visualizing, you are less likely to be inhibiting these intuitive brain processes with words.

An answer to the first question in the last exercise is to untwist the rope down its vertical axis and then join the two halves together. The answer to the second most of you will have got easily—a coin minted in 532 B.C. couldn't possibly have B.C. on it. And thirdly, to join the ropes, you tie the screwdriver to one of the ropes and set it swinging like a pendulum. You walk over to the other rope and catch the swinging rope. Attach them using the thumbtack and pliers.

Where language comes into its own is when the problems to be solved are logical and analytic rather than insight-hungry and in need of word-wary intuitive brain circuits. Try a logical, analytic problem that language *does* help solve. Talking yourself through this type of problem really does help, and how well you talk yourself through predicts how likely you are to solve it.

There are three playing cards from a normal deck lying face-down on the table. You know the following information about the three cards—each of these statements refers to the same three cards: To the left of a Queen, there's a Jack. To the left of a Spade, there's a Diamond. To the right of a Heart, there's a King. To the right of a King, there's a Spade.

Assign a suit to each of the three cards.

This type of problem is hard to solve without using language, reasoning, and logic. It demands *convergent* thinking, where there is a single answer. Insight problems, on the other hand, are made harder to solve when you try to use this type of thought process. These problems demand *divergent* thinking, where there can be more than one answer.

Fixed and Incubated

One reason we find it hard to come up with a new idea is that we get stuck on the most obvious aspect of a situation. Let's take the example of the puzzle of the prisoner's rope you tried earlier. Many people get stuck because they make the assumption that the rope has to be cut across its width rather than separated down its length. This is not something you are consciously aware of deciding or assuming. No, the conventional associations of "dividing" and "rope" involve cutting across the width—simply because 99 percent of the time ropes are divided this way. Mind-habit here binds you without you even realizing that this is what your brain has settled—no, let's call it *fixated*—on. Creative block is often caused by fixation, and it is hard to break through because we don't even realize that we have assumed anything. Cognitive scientists have managed deliberately to fixate people mentally and showed how this reduces their creativity.

Try to imagine and draw a completely new toy or game. Use any materials, objects, or shapes you like. For instance, you might think of soccer boots with an electronically controlled toe-part that makes sure you always kick straight. They do this because a radio receiver in the ball corrects its path according to the angle of the toecap device.

Here's a second task: try to imagine and draw a completely new creature.

How did you do? Let me ask you a question. Did your design for a toy or game involve a ball or an electronic device? If not, congratulations, because your brain was not trapped and fixated into using these objects by my suggestion. Now to the creatures. Did your creature have four legs, antennae, and/or a tail? No? Well, that's not too surprising, because I didn't give you an example where these features were present.

Research has shown that you can easily fixate people on a particular idea and so stifle the range of ideas they can come up with. For instance, if you have been given examples of new toys that contain balls and electronic devices, then you will be much more likely to "assume" that these should be part of your toy, even without realizing that you have been fixated on these elements. The same is true for the alien creatures. If I had given you several examples where the creatures had four legs, antennae, and a tail, you would have found it much harder to imagine a creature that did not have one or more of these features.

Let's try another example. You'll see three words below. Try to think of a fourth word that's individually linked—however remotely—to each of the three words. For instance, the words Type—Ghost—Story have the common associated word of Writer. Some of the words have related words in parentheses beneath them. Read, but then ignore, the words in parentheses.

Wheel	Electric	High	_____
(tire)	(cord)	(low)	
Ball	Storm	Man	_____
(soccer)	(tornado)	(boy)	
Family	Apple	House	_____
(mother)	(pie)	(home)	
Water	Pick	Skate	_____
Mouse	Blue	Cottage	_____
River	Note	Blood	_____

I'll give you the answers to these in a moment. These are items from a test of divergent, creative thinking called the Remote Associates Test. If you find it hard to solve some or all of these, forget about them for a while. Let them marinate in your mind. Then go back to them.

Steven Smith and Steven Blankenship of Texas A&M University found that it is much harder to solve the first three puzzles than the last three. This is because of the mind-habits switched on by the words in parentheses beneath the first three sets of words: these words create a false mental trail. They emphasize the conventional connotations and connections of these words, making it hard to think of all the other hundreds of possibilities there are.

Okay, let's give some answers (remember, this is a test of divergent thinking, so there may be more than one possible answer). For *wheel, electric,* and *high,* the answer is *chair*. The words *tire, cord,* and *low* take you well away from the *chair* connotations of *wheel, electric,* and *high* and make solving the problem harder.

The answer for *ball, storm,* and *man* is *snow*. The answer for *family, apple,* and *house* is *tree*. Again, on average, the trap-words make it harder to find the answers. It should, on the other hand, have been easier to get the words for the three pairs without mind-traps. In case you want to leave them incubating in your mind for a while, I'll give the answers to these in the notes section for this page at the back of the book.

You will be more likely to solve these problems—particularly where your mind has been lured into a particular connotation of the words—if you set them aside and stop thinking about them for a while. *Incubation* is a powerful antidote to creativity-blocking *fixation*. One major reason for this is that the fixation dissolves gradually over time, releasing its mind-numbing grip.

The Costs and Benefits of Convergence

It is not through nature's whimsy or bloody-mindedness that logical, convergent thought—and its chief lieutenant, language—narrows down our thought processes and switches off potentially attractive but logically irrelevant connotations. If you are trying to make a living as a hunter-gatherer in the early millennia of our evolution, it makes sense that the thought *deer* should sharpen up all the brain networks linked to deer—color, shape, movement—and turn down the volume on other brain circuits, such as those linked to upgrading your flint ax or plotting revenge on that fat-assed alpha-male who pushed you to the ground last night and stole your mate.

It's much the same if you are trying to pass your physics exam at school. Suppose you are chewing your pencil and tackling the question of which will fall faster in a vacuum under gravity: a feather or a lead weight. You will be much more likely to pass that exam if your brain has tuned up all the connections associated with gravity, force, and acceleration and has tuned out irrelevant associations to do with ostriches, trout, and exotic dancers.

So, whether it be as a hunter-gatherer or a sweating schoolchild, considerable advantages can accrue to those who are good at switching on those precise trembling webs of brain connections linked to the objects, actions, and concepts surrounding the most pressing task in hand—be it passing the physics exam or killing your dinner. But just as the girl who is top of the class in school isn't necessarily the headline-grabbing success in real life, so the deer-focused hunter-gatherer may in the long term lose his moll to the dreamy squirt who never managed to kill an animal in his life but invented the wheel.

In most Western societies, we give the big cheers to children who are good at the convergent, analytic, and word-based types of brain activity. They are the ones who come top of the class in our largely convergent, analytic, and word-based exams, but—take Einstein as an example—they may not be the ones who break the mold in later life, whether in science, music, technology, engineering, art, fashion, or business.

Essential though it is, logical, analytic thinking suits only certain types of problems. For more creative, divergent, and intuitive-insightful thought, words can act as glue rather than grease in the cogwheels of thought. Paradoxically, if your IQ is over 120, divergent and creative thought processes predict success in life better than the type of analytic thinking rated by IQ tests. In a study of the most eminent world political leaders, most of whom had IQs over 120, it was found that the higher the IQ, the relatively lower the level of eminence.

Great, so we get all our schools teaching creativity and divergent thinking and abandon all this oppressive convergent stuff? Hmm, hold on a minute. Several countries—the United States and Britain in particular—tried that in the 1960s and 1970s, and ended up with several generations of schoolchildren who could hardly read, count, spell, or write. No, let's get the balance straight—we need the logical, analytic, and language-based parts of our brain very badly indeed. This is not an either-or situation. We can't neglect one set of mental tools for the benefit of the other. What's more, real-life creativity demands logic and analysis as well as intuition and insight—and most of the time when we are being creative, we will be using both sets of brain circuits.

But we *are* neglecting the wordless world of imagery. Our children lose these native abilities in a word-based school system. It may be understandable. This insight and intuition business is pretty hard to pin down. After all, it's not word-based, so it's difficult to talk about it.

Can more creative, divergent thinking be taught? Can we rescue the withered faculties of nonword thinking?

Can Creativity Be Taught?

The answer to this question is a cautious and guarded yes. When schoolchildren are exposed to teaching aimed at creative, divergent thinking, they score better on tests of creativity. What we don't know yet is whether this translates into tangible achievements when they leave school.

These are the conclusions of the eminent American cognitive scientist Professor Ray Nickerson of Tufts University. He believes that all of us have a huge untapped creative potential, and he also argues that creativity is good for the quality of life of the individual as well as for society. Shaking off the leg-irons of ingrained mind-habit is a key element of unlocking this potential.

Using imagery, as we have seen, is one way of shaking off these shackles and accessing creativity, particularly where the problems you are facing are blue-sky, unfamiliar, and not amenable to deductive solutions. But problem *solving* may not be the key to creativity—finding what the problem *is* may be the key to creative breakthroughs in science, art, design, and technology. In other words, "What's the question?" may be an even more important challenge than "What's the answer?"

In the study of art students I mentioned earlier, the researchers followed up the students years later to see how dealers, critics, and gallery owners rated their creative success. And when the researchers looked back at how the front-runners had differed from the more modest achievers in the techniques and strategies they used as students, clear differences emerged.

The creative achievers tended to explore the subject they were painting much more—using all the senses. They felt it, sniffed it, held it to the light, moved it, weighed it, and so on. They also chopped and changed their approach a lot. The artists who had not made it creatively more than a decade later would have as students, on the other hand, tended to pick a subject, look at it for a relatively short time, and then plow ahead with the work without making many changes. It seemed as if the more creative artists were wrestling with establishing the right artistic question—they were finding the problem—for a longer and more tortuous period than the less creative people. They seemed to do this by deliberating, trying to shake their brains out of well-worn tracks of mental habit through engaging all the senses in exploring the object.

Children who are taught to question the question—i.e., not accept the immediate definition of a problem, but explore different ways of defining it—become more creative in the longer term, it seems. And imagery can help shake your mind out of the rut established by the obvious, in-your-face question.

The problem is that we tend to educate our children to meekly accept the problems posed for them and to solve them in quite conventional, analytic ways. There can be little doubt that this is at a cost of stunting some nonanalytic pathways where the challenge is to define the problem in new ways before it can be solved. Yet it doesn't take much to liberate young minds from these mind-habits. Take this research with young adolescent schoolchildren in London, for instance.

This study looked at the effects of just one lesson, once every two weeks, over a period of two years, begun with several classes of twelve-year-olds. Though this was a science lesson, it was just as much a lesson in thinking, and in particular in *finding* problems and questions, as in solving them. In other words, these lessons were more about teaching the children how to think than about teaching them specific scientific principles.

To give you an idea of the kind of training the children had, here is one type of exercise. These are not the specific examples, but the principle is the same. Think about these objects and concepts. Think of as many ways as possible in which you could classify or group them:

> coat, cloud, milk, bacteria, tree, committee, wind, ache, rough, anger, conflict, structure, stratosphere, core, rose, fear

To give an example, you could separate them into living versus nonliving, or abstract versus concrete. But there are many, many more possibilities. Find as many as you can and write them down.

The examples given to the children were more simple and concrete, but the principle was the same. What the children had to do here was something quite unlike what happens in most classes. They were used to being given a problem that had been defined by their teacher, and to solving it using methods that they had been taught by their teacher. Here, on the other hand, they had to *find* the problem—in this case one possible classification of a word, then another, and so on.

The children were not given abstract words like *committee* or *anger,* but were given objects and figures that nevertheless could be classified in many different ways, such as size, color, shape, angularity, etc. To do this the children had to master an abstract concept related to category, class, or group, but they also had to free themselves from the most obvious and familiar aspects of the objects in order to come up with many different categories.

Here we see the two mind-modes working in harmony. You can't easily grasp the concept of category without language. On the other hand, you won't find all the possible categories unless you tune down the language system from time to time to let the multisense imagery power of the brain have a chance to show its skills.

How did you manage on the categories test? How many did you get? You almost certainly didn't get them all. Look at them again, but this time use each mind-sense in turn to play with them.

Take the mind's eye first. VISUALIZE each of them in turn, and bring different pictures together. Imagine milk and committee and anger mixed up, for instance—but you choose which and how many and how to combine them.

Now take the mind's ear. Try to LISTEN to these different things, again combining them as you see fit.

What about the mind's nose? SMELL these things alone and in combination. Do the same for TASTE.

TOUCH them now, not just with your hands, but with your body too—imagine pushing them or moving through them.

Did any new dimensions spring to mind when you loosened the "cool web" and let the mind's nonword senses play over the concept? I'd be surprised if they didn't. To give one trivial example, when I used the mind's ear, I found that they could be classified into silent and nonsilent.

The teachers didn't specifically teach these London children to visualize. But as I hope you found out in the preceding exercise, it's likely that the kids who made full use of imagery would have come up with more possible dimensions than those who stuck with the word codes of the "cool web."

The dimensions exercise was only a small part of the training, though. The teachers would also give the children some objects and other materials and say, "What's the question here?" In other words, these kids had to use brain systems that had been badly underused in most of their education—namely, problem-*finding* skills.

The teachers also encouraged the twelve-year-olds to think about their own thinking. For instance, rather than just saying to themselves after a difficult problem, "Boy, that was difficult," they learned to ask themselves questions like, "What was it that was hard about that problem, and how did I overcome it?"

Remember, they did all this during one hour every two weeks of science experiments and teaching. The effects, however, spread far beyond science ability, into mathematics and even English exam grades. Two years after the end of the special course, the children who had been in these classes, on average, outstripped children who had not taken part, to a considerable degree.

Vision and Logic

This London schools study showed how a small amount of training in divergent thinking processes, added on to more conventional convergent training, produced better results even in more convergent, traditional types of school exams. This shows that we can't afford to be simplistic about the analytic-insight, convergent-divergent, language-imagery dichotomies. Very seldom can we get away with abandoning one in favor of the other. Life will throw up problems that demand both types of thinking—it's a question of being able to switch into and out of these often antagonistic mental modes. Because the "cool web of language" has such a grip on our mental life, however, it can be very difficult to suppress language and its mode of thinking in order to give free reign to the mind's eye. Brain damage can be one drastic way of releasing these capacities, but it is hardly practical as a general solution! I hope that in this chapter I've shown you some other ways in which this is possible.

There is, however, one other fundamental mental ingredient that underpins all creativity and problem solving—memory. Without the ability to store experiences, remember insights, and recall what we have read and heard, we are crippled mentally. The mind's eye is crucial for memory. Let's turn to that now, in the next chapter.

6

The Landscapes of Memory

Look at the butterfly. Let your mind roam through your past and think of some event in your life that a butterfly reminds you of. Something specific, not general. A certain place, a particular day. Pull out that moment from your memory. If you can't, try with the word lake*—see what that evokes. But notice how you don't just remember the butterfly or the lake. When you remember an episode from the past, then sensations and feelings linked to the episode will flood back too.*

Can you remember your first day at school? What about a wedding—your own, or someone else's? The birth of a child? Try to imagine yourself back to a particular moment or scene in your past linked to these. What strikes you about these memories? Do you notice how the flimflam of experience clings to the memory: a certain light in the sky, a mood—intense happiness or fear, a movement, an object, a fragment of music, a taste, an ache, a thought?

A memory. A particular lake in Glasgow, Scotland. Saturday morning, early summer—June maybe. The sailing dinghy heeling over against the fresh wind, the lake water dark except where the sun glints blindingly on a wave crest. The hollow thump-thump *of the waves against the hull, and the chill spray against my fourteen-year-old face. My feet tight in the toe straps and the pleasant strain in my thigh and back muscles as I stretch out against the wind, sail rope tight around my right hand, tiller in my left. And, as I remember, the most acute sense of happiness—the feeling of "Yes, this is it!" Yet with that intense feeling went a remembered thought—the thought that this intense happiness would end, that it was transitory.*

Memories of Self

Think of these memorable moments in your life—the birth of a child, the death of a parent, the day you met your partner, the day you got that job, the first day at school. Now imagine that you can't remember them. Imagine that your memory for much of your life has been erased. Yes, you have learned that your first child was born in 1994, but you remember it as a fact, not an episode. In other words, imagine that you can only remember personal experiences in much the same way as you can remember what is the capital of Spain.

Try that for a moment—remember the capital of Spain. Was this dipping into memory anything like when you remembered some personal episode in your past? Not really—because recalling that Madrid is the capital of Spain brings with it no flimflam of experience. You won't remember where you were when you first learned this fact, and there won't be any sensations, feelings, or thoughts linked to this type of remembering. This is because knowing a fact about Spain depends on a different memory system of the brain than remembering—for instance—the last time you ate an ice cream. The first is called *semantic memory*, while the latter is called *episodic memory*—you read a little about this in Chapter 2. And where the brain has different systems to control different types of mental activity, then it is tragically possible for one of these to be damaged or knocked out while the other survives. So it is with semantic and episodic memory.

John's daughter was two years of age when he was catapulted from his bicycle by a car into a six-day coma. The impact damaged John's brain, and this demolished his memory for past episodes from his life. Tragically, John had lost that most precious of human gifts—the ability to immerse himself in memories of the past and reexperience them with all the multisensory richness and imagery the mind can supply. Take the example of my remembered morning sailing as a teenager. If the same parts of my brain had been damaged as were in John's case, then I wouldn't be able to relive that vivid morning in my memory. Instead of feeling myself transported back all these decades, albeit in an imperfect way dulled by time, I would have at best remembered the morning's sailing as a flat, bald fact. It would have been as if I had read it in a book rather than experiencing it myself.

Imagine that your most intimate memories were stripped of the personal quality of *you* having experienced them. Imagine the memory of a first child being born becoming equivalent to the memory of the fact that swallows fly south at the end of summer. This happened to John, who was studied intensively by Brian Levine, Endel Tulving, and their colleagues at the Rotman Research Institute in Toronto.

Five years after the accident, John could remember only a handful of fragmented images from his entire life before the accident. And these were like images from very early childhood—fragmented, disconnected, and not located in any particular time or place. To be brutally brief about it, John had lost his own personal past.

But what about his more recent memories since the accident? After all, his wife was heavily pregnant when he had the accident, and gave birth to their second child while he was in the hospital. John could remember a list of words as well as you or I could—so surely his more recent memories would not be stripped of the sense of "I was there," with all the images and sensations that go with this. No—even these more recent memories were remote to him, and cruelly impersonal, as if he had read about them happening to someone else. How, then, could his memory for new factual information be normal? The answer seemed to be that he was now remembering information in a different way from before. When learning, his brain no longer stores the flimflam of context, but rather homes in on the raw factualness of the precise fact to be remembered. It's as if his memory were stripped down to the Madrid, Spain, type of learning and could no longer pull out the "I remember sailing on the lake" type of memory, with all the sensations and images that go with it.

We might be able to enrich our own personal memories by using imagery as an extra source of context—more useful flies on the windshield of memory. More of this later.

Here are some events you will be familiar with. Think of each of them, and try to answer this question: Do you just "know" the fact, or do you actually remember something about what you saw, felt, heard, or thought when you first learned it? Let's call the first "I just know" and the second "I remember something about the occasion when I learned it." Write K for the first and R for the second, opposite the examples below:

President Kennedy was assassinated.

Neil Armstrong was the first man on the moon.

Yuri Gagarin was the first man in space.

There was an explosion at Chernobyl Nuclear Power Station.

The Challenger Space Shuttle blew up.

John Lennon was shot in New York.

The first nuclear bomb to be used in war was dropped on Hiroshima.

A Pan Am jumbo jet exploded over Lockerbie, Scotland.

Which of these you simply know and which you truly remember learning depends a lot on your age. If you are eighteen, then you could not possibly remember John F. Kennedy's assassination. But even if you were around for a particular event—John Lennon's death, for instance—it may not have been particularly shocking or important for you, and so your brain may not have stored the surrounding flimflam. Personally, I only "know" about Hiroshima, as I wasn't around to learn about it when it was news. And while I vaguely remember about John Lennon's death, I don't have the "I was there" memory that I do, for instance, about Chernobyl or Lockerbie.

When John—the Canadian man who lost his personal memories—was given a remember-know test for a set of definitions he had to learn, he learned them well, but was much less likely to remember anything about the circumstances in which he learned them. That was how his brain had adjusted to the damage, but at what a cost: stripped-down memories without the accumulated stick-ons of context.

Wait a minute. How can these—the peal of laughter, the glint of sunlight, the quirky expression, the glimpsed cloud formation—be written off as flimflam and stick-ons? These are the very fabric of our cherished personal memories. And when it comes to unlocking these glimpses from our personal autobiographies, there is nothing so powerful as a visual image. Research shows that when we remember personal events from our own experience, visual imagery is almost always involved. What's more, the more easily people can create an image of a word, the faster the memory will be evoked and the more specific and detailed it will be.

Try to think of a personal experience/memory that is triggered by the following words:

Law
Effort
Situation
Duty

Now try these:

Grass
Factory
Baby
Sea

If you are like the people who took part in a study by Mark Williams and his colleagues at the University of Wales at Bangor, then you would have come up with much more detailed and specific memories for the second set of words than for the first. This is because *grass, factory, baby,* and *sea* are much more "imageable" than words like *duty* and *effort,* even though they are used equally often in everyday speech and writing.

But is it the visual images of words that are most potent at triggering personal, autobiographical memories? Couldn't smell, sound, and touch be just as effective, if not more so? Try yourself with these words that are linked to other senses.

Cheese	Coffee	Smoke	(smell)
Ice	Sponge	Wool	(touch)
Laughter	Choir	Snore	(sound)
Spade	Racquet	Ax	(bodily movement)
Cloud	Fire	Mountain	(vision)

Maybe you found that one of these senses was better than others at triggering vivid memories from your past. In the study, however, it was found that, on average, the words that trigger visual images aroused the most vivid personal memories. The second most potent sense was sound—words like *thunder, choir,* and *snore* were a close second to vision in their memory-triggering potency.

The Machinery of Self

The blow to John's head damaged his brain on the underside of the frontal lobe on the right side. This area is just above your right eye. The blow also severed the neural wiring connecting this part of the brain with the front part of the temporal lobe on the right side. This is slightly farther back on the outside of the brain, on the right side, between your eye and your ear. This "wiring" consists of long strands of white fibers connecting brain cells in the cortex, and the particular one damaged in John's case is called the *uncinate fasciculus*.

Now the front part of the temporal lobe (literally—at your temple) on the right side seems to be a key part of the brain's storehouse for personal memories of past episodes. This is where your autobiography of experiences—the ones you remembered in the exercises above—is largely stored. People who suffer damage to this part of the brain often—like John—lose part or all of their autobiography, the part that, as we have seen, is so heavily linked to images and the mind's eye.

In my professional life, I have seen several unfortunate people like this. One young woman, following a head injury, lost, as did John, her entire personal, subjective memory for her life prior to the age of twenty-five: She couldn't remember school, friends, her twenty-first birthday party, or any of the key events that are so precious to everyone. Through conversations with her family, and by looking at photo albums, she relearned her own history to a small extent, but it was not the reliving of personally experienced episodes with all the sights, sounds, and cornucopia of personal detail that go with episodic memory. No, this was a flat, third-person relearning of a life no longer subjectively experienced.

I remember another young man—in his early thirties—a devoted father who could remember nothing of the birth of his two young children and had lost all his personal memories of their infancy and early years. He couldn't remember his wedding, and only "knew" about how he met his wife by what she had told him subsequently.

But in John's case, the brain storehouse for personal memories—the front part of the right temporal lobe—didn't seem to be particularly badly damaged. What *was* damaged was the connection between this part of the brain and the frontal lobe on the right side. And it seems that the right frontal lobe is also a key part of the machinery of self. The right frontal lobe, at its underside, seems to be a kind of stationmaster for the trains of autobiographical memory. It sends signals to the storehouse of personal episodes farther back in the temporal lobe, helping reawaken acutely subjective memories, like mine as a teenager sailing on a lake.

As you were remembering incidents from your past in the exercises above, the right frontal lobe of your brain would have been particularly active. Take a moment to remember the last wedding you were at. As you do this, the brain cells in the right frontal lobe, just above your right eye, are firing more actively, and your brain is sending more blood to fuel this activity. If your head were in a brain scanner, we would be able to see this extra brain activity as your frontal lobe sent patterned signals back to the storehouse in the temporal lobe to arouse the triggered memory.

The triggers can be external—the glimpse of a picture, a particular smell—or internal, such as when you try to remember the last wedding you attended. It is when you search your memory in this latter way that the right frontal lobe is essential. Brain cells in the frontal lobe have been shown firing, triggering particular memory-related cells in the memory storehouse in the temporal lobe, in response to such an internal request.

In John's case, however, both the frontal lobe and its connections with the temporal lobe were damaged, so signals could not be sent back to his memory storehouse in order to reawaken personal memories. He did have some isolated fragments of personal, subjective memory of his past, and it may be that chance reminders from the external world triggered them. What he couldn't do, though, was to summon up these memories at will.

Endel Tulving thinks that the ability to remember these personal episodes is one manifestation of a self-knowing awareness generated by our brains that gives us the thread of continuity back into our pasts and forward into the future. It allows us to have the sense of being a continuous entity over time. Because of this network of regions in the right hemisphere of our brains, Tulving argues, we know that the self that experienced those events years before is the same self that is reliving them now.

And this reliving entails the brain trying to reassemble the circumstances of the original event. How else does it reconstruct one moment of experience out of the trillion upon trillion that have impinged on it in a lifetime? It does so by reassembling the images, sensations, and feelings that littered the stage for that one attentive moment. How often was I fourteen, simultaneously ecstatic and morbid, sailing on a Saturday morning in June on that lake, with the wind making just that pattern of waves, and the light deflected just so by the random clouds of the moment? Only once. And that is why I could reassemble it out of the near infinity of recorded experience. But without images projected onto the mind's eye, I could barely have done it.

Learning Visualized

Words can corrode certain types of thinking, and they can cloud the mind's eye—I showed you that in the last few chapters. Robert Graves, of course, knew this long before I did. The same can be true for memory and other realms of sensation.

Are you a wine connoisseur, or are you an insensate amateur oblivious to the finer nuances of the grape? Do friends who fancy themselves as cognoscenti after a couple of classes browbeat you with loquacious reflections on the audacious gooseberries, oak-tainted butter, and mellow jam that fill their mouths? Do you tend to stay mum and just drink the stuff? If you do, then take heart. Unless they really are trained wine experts—and you know how few of them there are around—if they talk while they taste, they will surely snuff out their brain's ability to tell wines apart.

The test in this research was simple. Trained experts and would-be noses tasted a red wine, and then either talked about the taste, or spoke about something irrelevant. They were then given three different wines, and they had to say which was the one they had tasted a little while before. While the experts did well no matter whether they had talked about the wine or about their gardens, the amateurs—i.e., the ones like you and me—couldn't remember the first wine properly if they had been talking about its taste and quality. If they were talking about something else, on the other hand, then they could remember which one they had tasted. So the next time you are irritated by someone's amateur musings about the quality of your wine, just tell them that unless they shut up, they might as well be drinking sugary antifreeze for all their brains will be able to tell!

Yes, if we rely too much on the "cool web of language" in our contact with the world, we really do cut ourselves off from certain more immediate experiences and the skills and benefits they can produce. But it isn't just in memory for taste that it is important to free up and use our mind's eye, mind's palate, etc. It's also true for that most essential of day-to-day survival skills—learning.

Read this list of words twice, and try to remember them. When you have read them, close the book and see how many you can remember:

dog, chair, hat, cloud, car, umbrella, pen, hen

Now read this second list. This time, try to picture the object in your mind as you read each word. Read the list just once, but make sure you have created an image of each object in your mind's eye. As with the first list, see how many you can remember:

table, glove, pencil, plane, cat, eagle, briefcase, moon

You should have found that it was easier to remember the second list, though there can be exceptions—some people find it hard to create visual images and so don't benefit from this. In general, however, we are more likely to remember *dog* if we see it as a picture than if we see or hear it as a word.

A Picture Is Worth a Thousand Words

One explanation for this is that seeing the picture gives you two brain activations for the price of one. Not only does the brain's language system store the concept 'dog,' but the brain's visual image system (and maybe also its touch system) also stores a copy of 'dog' via the mind's eye.

You tend to remember things when the frontal lobes of your brain are active and firing at the time of learning. That's why, if you are switched off and spaced out, you will probably not be able to recall events or facts that you might otherwise have learned. This is why teachers are so concerned for children in their classes to be attentive—attention is controlled by the frontal lobes, and if you don't attend, you don't learn and remember.

This is also why teachers and parents are not at all eager for children to listen to music or watch television while they are studying. Unless it is really dull background music that can readily be ignored, it is inevitable that attention is divided between the homework and the music or TV. And when attention is divided like this, the memory-related brain activity in the frontal lobe drops. So the next time you see someone trying to study in front of the TV, it might be worth suggesting that they could spend half the time studying without the TV and remember just as much—because the distracting TV will be turning down the volume of brain-cell activity in their frontal lobes.

We know about the importance of the frontal lobes in memory because of brain-imaging studies where volunteers were asked to learn words or pictures while their brain activity was recorded. Later on, they were asked to remember these words or pictures, and, of course, they only remembered some of them. The researchers then looked at what was happening in the brains when they were learning things they later forgot versus things they later remembered. The difference was the firing of the brain cells in the front part of the brain. Memory for unknown faces or patterns that you can't easily put into words tends to be linked to activity in the right frontal lobe of the brain, while word memory is linked to activity in the left frontal lobe. But when the researchers looked at the kind of exercise you just did—learning words like *dog* and *cloud*—they found that there was activity in *both* frontal lobes. In other words, when you learn words that can also trigger visual or other images, you get twice the number of brain areas called into action. This may explain why you remember the picture of the dog better than the word *dog*.

"You Know . . . My Memory Just Isn't What It Used to Be . . ."

Everyone's memory gets worse with age, but visual memory holds up better than word memory. This is another example of the strange cycle of the human life—bald, shaky on the legs, toothless, and quirky at both the beginning and the end. And so too, it seems, for visual memory.

Some of you reading this book will be well used to using your mind's eye in thought and memory. But many of you will be wordy people who have let your mind's eye grow sluggish. This is a pity, particularly as you hit the late forties and fifties when memory begins to let you down. It will let you down a lot less if you start using the mind's eye and the multisensory, wordless images it can generate.

It's very interesting to compare the brains of older and younger people as they try to memorize things. Take this list of words: bread, couch, carrot, milk, fish, apple, chair, shelf, table. Take a few seconds to try to memorize these words. Give this to a group of twenty-year-olds to memorize, and their brains will show a healthy surge of activity on the left side of their frontal lobe as well as in the main memory centers of the brain deep in the middle, in a region called the hippocampus.

What happens in the brains of seventy-year-olds? Well, for a start they won't be able to memorize long lists of words nearly as well as the twenty-year-olds. But why? A peek into what their brains are up to while they are memorizing gives us a clue. The seventy-year-olds don't switch on the left frontal lobe nearly as much as the younger volunteers, and this is a likely reason why they don't remember as well.

Let's take another look at the list of words that you tried to memorize. Did you sort them out into categories to help remember them? In other words, did you sort them in your mind into "furniture" (couch, chair, shelf, table) and "food" (bread, carrot, milk, fish, apple)? If you did, you would have found it much easier to remember them than if you didn't. Older people, it seems, for some reason don't do this kind of mental sorting when they are trying to remember things—they don't actively sift, sort, and categorize in the way younger people are prone to do. And where does such sifting and sorting happen in the brain? You've guessed it—in the frontal lobes.

So what happens when you give the older volunteers a hint and tell them to sort out the words into categories when they are memorizing them? Sure enough, their memories improve. Not only that, however—they also show reasonably normal youthful activity in the left frontal lobe of their brains as they memorize.

One of the reasons our memories let us down as we get older, then, is that for whatever reason we don't attack the information with the same brain vigor that we applied when we were young. Laziness? Perhaps, but it might also be that we have got out of the habit of learning new things, and the brain circuits—in the left frontal lobe, for instance—have become lazy through disuse. This is just a hypothesis, but in my book *Mind Sculpture,* I showed that brain cells and brain connections can be generated by the right kind of mental stimulation. This might be part of the story—certainly not the whole story—as to what is going on as our memories worsen with age.

What can we do to keep our memories in good shape as we get older? Well, the obvious way to do this is to learn to "attack" the things you have to remember by sifting, sorting, and linking them to other things you already know. This is called *depth of encoding,* a principle discovered by the eminent cognitive psychologist Fergus Craik, of the University of Toronto. He and his colleagues have conclusively shown that the more you mentally process things you have to remember, the better you remember them. This is very likely in part because mental processing of this type activates the frontal lobes, which in turn strengthens connections between different items stored in your memory banks in the temporal lobes.

Read this list of words, deciding whether each word begins

with a vowel or a consonant. Read only once, close your eyes, and see how many you can remember:

> wall, ostrich, field, bell, slug, soldier, soil, acre, girl, tree

How many did you get right? If you got them all after just one reading, that is pretty good indeed. Most of you will have remembered only some of them.

Now read this second list of words, this time deciding whether each word is living or nonliving. Again, read just once and try to memorize them:

> floor, sky, turkey, tower, pilot, worm, grass, man, mile, flower

How many did you get right this time?

Many of you will have remembered more of the second list than the first. This is because making the judgment living versus nonliving forced you to do more mental processing of the meanings of the words than in the first list, where you only had to make a superficial judgment as to whether the words began with vowels or consonants. Forcing your brain to penetrate into the meanings of the words to be learned automatically made a stronger memory trace for each word.

Making mental images of the things you are learning is a particularly powerful way of burrowing memories into the brain so that they are less easily forgotten. For older people this is particularly good news, though the principles apply equally to people of any age. Image-based memory seems to hold up better with age than does word-based memory, and this makes visual imagery a particularly promising way of boosting memory for older people.

People differ in how much they tend to use the mind's eye

versus thinking in words. Answer these questions. Do you often use the mind's eye to solve problems with mental pictures? Can you easily see objects moving in your mind's eye? Can you do arithmetic by imagining the numbers chalked on a blackboard? Do you find it hard to make a mental picture of anything? Do you prefer problems where you need to use words (e.g., crosswords) to ones where you have to use mental images (e.g., whether that wardrobe will fit into the bedroom)? Is most of your thinking verbal, as if you are talking to yourself in your head?

If you answered yes to the first three and no to the last three, you are probably more of a visualizer than a verbalizer. If, on the other hand, you answered no to the first three and yes to the last three, you are probably more of a verbalizer. Many people will be somewhere in the middle.

Are You a Visualizer or a Verbalizer?

It's interesting to ask your family, friends, and colleagues how they place themselves in that dimension. I recently asked a colleague of mine, an eminent physicist, whether he thought mainly in words or in pictures, and he said without hesitation—"pictures mainly." An equally eminent philosopher colleague was equally unhesitating when he responded that he thought "almost entirely in words."

Einstein didn't have much time for words—he said that language didn't play a big part in his thinking, and that his education really emphasized imagery and nonverbal thinking. This distinction between visualizers and verbalizers—remember, it is a continuum and not an either-or situation—was made by another of the pioneers of research into mental imagery, Alan Paivio. The questions above are adapted from a questionnaire he developed to measure how people differed in this dimension. He discovered that people who tended to answer yes to questions like the first three in the exercise and no to the last three were better than their wordy colleagues at solving non-word-based puzzles that force you to use the mind's eye. It's not surprising, really, that the professor of physics said he was mainly a visualizer, and the philosopher that he was a verbalizer—many problems of the physical world can only be tackled by abstract, nonverbal thought, including mathematical thought.

The visualizers among you are probably going to find it easier to use the mind's eye to boost your memory, but it is the verbalizers among you who would benefit more by training the mind's eye and increasingly using mental images. As you saw earlier in the book, it is possible to teach yourself to be a better visualizer.

If you are a visualizer, you will be better at remembering pictures and words that easily spark off images—e.g., rose or tree—than the verbalizers are. Whether you are a verbalizer or a visualizer, if you take the trouble to use your mind's eye to picture what you are learning, this will in general boost your memory for what you have learned. One caution here, however: sometimes it can take time to create mental images, particularly if you are more of a verbalizer, and this can slow down learning. You have to be choosy as to where and when to use this strategy, at least until you are expert at it.

Children as young as seven can be divided into visualizers and verbalizers—again with many being somewhere in the middle, able to use both as the occasion demands. When you give these children stories or text to read, you get startling differences in memory. Children who are visualizers are much better at remembering words and sentences that are easily visualizable. For instance, they will be much better able than verbalizer children to remember the sentence "The stars on the European flag are white." Children who are more word-orientated, on the other hand, will be better able than their visualizing peers to remember the sentence "People use salt more than pepper." This is because the abstract concept "use more" is very hard to visualize and is better captured by the abstraction of language. The stars on the European flag, on the other hand, are easy to visualize, and the concept "on" is concrete and easy to picture.

You can see from this that it is not the case that imagery is "good" and verbalizing "bad." They are both incredibly powerful tools, and we badly need both to use our brains to their full extent. The problem lies in neglecting one in favor of the other.

Here are two short passages of prose. Read them to yourself silently, and time how long it takes to read each one.

1. The cows' udders swung like heavy bells as they plodded, slip-slopping through the steaming newly fallen dung of the leader. Their breath rose in the still, frosty air, smearing the crisp outline of the rising sun. The last cow reared and kicked as a large brown rat darted between its legs.

2. The sheep were more common in the relatively barren uplands, though the economics of sheep farming were becoming marginal and these pastures were gradually returning to their native state. Though the visitors from the nearby city liked this, it was a tragedy for the farmers who had lost their livelihood.

Both these passages are fifty words long. How long did it take you to read each? It took me around eleven seconds to read the first and eight seconds to read the second. This was in spite of the fact that the second one had longer words.

As an adult, if you are a visualizer, you would be more likely to be slower on the cow passage than the sheep one. If you are a verbalizer, the difference would be much less, and maybe even reversed. In general, visualizers take up to 25 percent longer to read prose like the first passage than verbalizers do, while on more abstract passages like the second one, they read at about the same speed. But this isn't a disadvantage for the visualizers, because adult visualizers remember the high-imagery material better than the verbalizers do, but are no worse on the abstract passage. This is probably because when the visualizers read, the information is stored more widely in the brain, both in the language circuits and in the mind's eye.

Women report, on average, that they can create more vivid mental images than men. Men, on the other hand, report that they can better control and manipulate mental images in the mind's eye. Earlier in the book I showed you that picturelike images full of vivid shapes and colors are almost certainly generated in a different part of the brain from where shape- and space-type images are created. If you remember, even people who have been blind from birth can create the latter kind of abstract spatial mental images, but they can't create the more pictorial type of image. The male-female difference is just a statistical trend—there are many women who are better at spatial imagery than men. Nevertheless, it seems that, on average, a woman would be better able to imagine how a new sofa might look in the decor of the living room back home, while her partner would be better able to picture where it would fit in the spatial layout of the room.

It may be that the slightly different organization of men's and women's brains explains why there are slight sex differences in the mind's eye. But these are trivial compared to the dramatic differences in word versus image capacities that can arise because of genetic disorders. Williams syndrome is a genetic disorder on chromosome 7 affecting about 1 in every 20,000 boys and girls born. Autism is also a disabling condition that is highly inheritable and may be linked to a disorder of the genes in chromosome 13. Autism afflicts boys twice as frequently as girls.

Williams syndrome children are extremely verbal, very socially skilled, and tend to be self-possessed, articulate, and often witty. They tend to be very interested in social relationships and initiate contact with people often in an inappropriately overfriendly way. It is only after a while of talking with one of these individuals that the gap in their intellectual abilities becomes apparent. This is because their nonverbal intelligence is severely reduced, and their precocious apparent intelligence is very easily punctured. In other words, they may know a lot of quite big words for their age, but they are very limited in understanding the abstract concepts behind the language.

Williams syndrome children, then, have a big difficulty when the "cool web of language" is shaken, and flounder with the simplest type of nonverbal thinking. And as you might expect, these children are verbalizers rather than visualizers, and they do much better on verbal memory tests than on visuospatial memory tests.

Autistic children, on the other hand, tend to have great difficulties with language and with social relationships, while often showing relatively good visuospatial abilities. In fact, a few autistic children like Nadia, who drew the horse shown in Chapter 2, have superdeveloped visual and visuospatial abilities. For them the "cool web of language" is indeed a spider's web in which they become trapped. For the Williams syndrome children, on the other hand, language is a mental life raft.

Seduced by the Mind's Eye—Misremembering the Past

From time to time, people confess to crimes they did not commit. Sometimes the false confession only emerges decades after the crime, or—tragically—after the confessor has died. Why on earth do people confess to crimes they are innocent of?

There are many different reasons, but one important factor has to do with how our brains retrieve memories from our autobiographical memory stores. I told you earlier that remembering incidents from your own past draws heavily on visual imagery. The problem is, however, that visual images are often seductively convincing and real—even when the events imagined did not happen.

Try to remember if this incident happened in your childhood: It's after school, and you're playing at home. There's a strange noise outside, and you run to the door to see what made the noise. As you run, your feet catch on something and you fall.

How confident are you that an incident like this actually happened in your childhood? Give your confidence a rating between 1 (not sure if it happened) and 4 (certain that it happened—can remember it clearly).

Now take this second incident: You are falling, and as you reach out to support yourself, your hand goes through a pane of glass. Your hand gets cut and there's blood.

Again, how confident are you that such an event happened to you in your own childhood? Give a rating between 1 and 4.

Now take the first incident. Close your eyes and try to picture the situation. Try to make a mental picture of your home, try to imagine the scene vividly, and try to "feel" yourself running and falling. Take about a minute to try to re-create the visual image of this, and also the bodily feelings of running and falling.

Do the same now for the second event—the cut hand on the broken window. Imagine the scene as vividly as you can in your mind's eye. Try to feel the pain and shock of cutting your hand—try to imagine the sound of the breaking glass.

Now rate again how confident you are that these events happened, again on the 1–4 scale. Has it changed?

Most of you won't have changed your ratings of confidence that these events occurred, because there was such a short space of time and you will have remembered the first rating you gave.

When Marryanne Garry and her colleagues in Indiana carried out a study similar to this, however, they left a two-week gap between the first time people were asked to recall these events and the second. What's more, they asked the volunteers about eight such incidents, not just two. For half the incidents, chosen at random by the researchers, the volunteers had to try to mentally picture the incident in question at the second visit. They then gave a confidence rating about how likely it was that the incident happened. The researchers found that events that were re-created in the mind's eye on the second visit two weeks later were rated as more likely to have really happened in the person's life than the events that the subjects simply had to think about without creating a mental picture.

Now this couldn't have been because the events actually did happen. Why? Because the mentally pictured events were chosen at random by the researchers, and it would have been statistically highly unlikely that these randomly chosen events would have chanced to be the ones more likely to have happened.

Another piece of research confirms that the mind's eye can be too persuasive and can cause you to misremember the past. In this study, the researchers interviewed the parents of young adults to get information about events that really happened in their children's lives. Over three interviews, the researchers repeatedly asked the young people to remember these events. Sneakily, however, they also threw in another event that hadn't happened to them—this was a fictitious happening that the researchers made up. But like the real episodes from their past, the volunteers had to try to remember this imaginary event.

Here's where visual imagery comes in: Half of the volunteers were asked to mentally picture the events they had to remember—including the made-up episode. The others were simply asked to sit quietly and think about these happenings, without being asked to use the mind's eye to create mental pictures of them. When people used their mind's eye in this way, they were much more likely to suddenly "remember" the fictitious event as actually having happened to them. The visual imagery and the repeated recall of these hazy events crystallized into a self-deceiving sense of "Yes, I remember that happening to me."

You can probably appreciate now how it is that some people end up confessing to crimes they did not commit, particularly if they are young and mentally vulnerable. Police interviewers can sometimes try to extract confessions by requiring people to picture the scene and circumstances of the crime. When they do so again and again, it is entirely possible—perhaps even likely in a vulnerable person—to create the false memory in the person's mind that they did do the crime.

Young children are even more vulnerable to the unintentional planting of false memories in their minds. I showed you in Chapter 2 how four-year-olds weren't easily able to distinguish between being told a story about packing a picnic basket and actually having packed it themselves. Six- and eight-year-olds—with their better-developed frontal lobes—were much better able to distinguish fact from fiction.

This is a terrible headache for police and child-care workers who have to deal with the problem of child abuse. For decades children were not believed, and abusers could continue without hindrance. Now children are much more likely to be believed, but with this has come the problem of distinguishing fact from fiction in these young memories.

This is why it is so important to have corroborative evidence from the behavior of children and from medical examination. It is also why it is so important that the interviewing of children be very carefully conducted to avoid through repeated questioning the risk of implanting memories. Nevertheless, because our memories are vulnerable to self-deception—particularly where we use mental imagery for distant events—people must be vigilant and try to minimize the risk of false accusations, and false confessions arising from false memories.

The Brotherhood of Memory

If you use the mind's eye more, you can improve your memory. This fact should be no surprise given what you've read in this chapter. After all, when you use visual imagery, auditory imagery, or "visualize" in any of the other senses, you switch on more of your brain than when you confine yourself to word-based remembering. This is crudely put, but it is true. And the more areas of your brain that are involved in the capturing of one of these fleeting moments of experience, the more likely it is that you will be able to trace your way back through the immensely complex web in which fragments of memory are knitted.

In my book *Mind Sculpture,* I explained how memories can be embroidered into a network of up to 100,000,000,000 brain cells. On average, each of these hundred billion brain cells is connected a thousand times with other neurons, making a total of 100,000,000,000,000 connections. These hundred trillion cell meeting-points in a human brain exceed the number of stars in our galaxy. The meeting point between two cells is called a *synapse*.

A brain cell is shaped a bit like an onion, with a roundish middle, a single long shoot at one end, and lots of thinner rootlike fibers sticking out the other. An onion sucks up nutrients from the ground through its roots, processes them in the onion body, and sends the results up into the sprouting shoot, and brain cells work a bit like this too.

The cluster of thin fibers converging on the brain cell corresponding to the onion's roots are called *dendrites*. These, like the roots of the onion, suck up nutrition into the brain cell. But the particular "nutrition" that they bring consists of electrochemical impulses from other brain cells. The tendrils from these other cells itch and nag at our cell's surface, trying to annoy it into action. Others have the opposite effect and tend to inhibit the cell they link to. Whether or not a brain cell fires depends upon the final arithmetic of all the combined hectoring of these inputs to our cell—all the "*go go!*" inputs together minus all the "*stop stop!*" inputs. And once this arithmetic goes above a particular level, the brain cell fires, sending an electrochemical impulse shooting up the equivalent of the onion's single green shoot.

This green shoot on the brain cell is called its *axon*. A cell has a single axon, which is its only channel out into the rest of the brain. Axons can range in length from a tenth of a millimeter to two meters. When the brain cell is coaxed into sending a signal down its axon, it does so in a single blurting pulse, not with a constant trickle. This pulse lasts about a thousandth of a second and travels at a speed of anything between two and two hundred miles per hour. It travels down the axon, where it causes an itch at the point of contact with the dendrites of another brain cell. That point of contact is the synapse. This response continues through the trembling web of neurons connected by synapses; and so a chain reaction occurs, with cells firing off in their hundreds, thousands, or millions across the three-dimensional net.

At this very moment, as you read this sentence, exactly this cascade of brain cells firing is happening in your brain across these all-important junctions—the synapses.

Earlier in this chapter, I asked you to read two short passages of prose. Can you remember any of either of them? One was about cows, the other about sheep. Try to remember as much as you can of each.

You in all likelihood remembered some of the passages—probably more about the highly visualizable cows than about the less imageable sheep.

The reason you remembered anything at all was because your brain managed to connect the pattern of brain-cell firing linked to the stories themselves with the pattern of brain-cell firing that stored the *context* of this learning. For you to have had any idea what I was talking about when I asked you to remember the exercise earlier in the chapter, you had to re-create from memory the situation where and when you were sitting reading this chapter, be it minutes, hours, or days ago.

If your brain had stored none of this context, my request that you remember the passages would make no sense. What I was doing was giving your brain a few clues about the context ("Earlier in this chapter, I asked you to read"), and with these fragments your brain managed to re-create the moment when you were reading these two passages of prose. Only when it had constructed this context could your brain then pick out the information that was gleaned in that context—the swinging udders of the cows and the darting rat, in the first passage. These words and images triggered a distinct pattern of firing in millions of onionlike brain cells throughout your brain—probably many more brain cells than the sheep story triggered.

At the same time, the context (me reading a book on the bus half an hour ago, for instance) triggered another unique pattern of brain-cell firing. So we had two vast pulses of activation spreading through your brain as you read the two stories—one for their content, and another for the context in which you were reading them.

The key to memory—and to you being able to improve your memory—hangs on the brain's central trick, a trick that allows the brain to bind together these two sets of memories. As I just said, a brain cell fires when it gets enough of a push from the axons of other brain cells itching and nagging at its surface. As this happens, cells that don't have all that much to do with each other end up firing off at more or less the same time. This is not because they have always been wired up together, but simply because both happened to be triggered by the same cascade of activity in the brain—i.e., the surge of electrical activity generated by you reading the stories, as well as the activity triggered by the context.

It's a bit like being stuck in a delayed train with someone: at first you may not speak, but after an hour or two you will both be groaning and complaining together. A similar thing happens to brain cells. After a few repetitions of firing together, they tend to team up. When two connected neurons are triggered at the same time on several occasions, the cells and synapses between them change chemically such that when one now fires, it will have a much bigger punch in triggering the other. In other words, they become partners and in future will fire off in tandem much more readily than before. This is called *Hebbian learning,* after the Canadian psychologist Donald Hebb, and the chemical change in cells and synapses is called *long-term potentiation* (LTP).

So as you read the stories, the brain cells that registered the words and images of the stories themselves fired at the same time as the brain cells that registered the context of you reading them. And, hey presto, because they were all firing at the same time, you got these two waves of brain-cell activity connected on the basis that cells that fire together, wire together. When I then asked you later to remember the stories, the context brain cells that I triggered by my request nudged awake their new sister brain cells registering the story content of udders, cows, and sheep.

As we live our lives, experience gradually remolds us. Connections are made in the brain, and connections are broken. We learn and we forget. Anger transmutes to guilt, affection to resentment, despair to hope. Some memories will not be forgotten: the smell of new-baked bread from the bakery in that small French town where you had that wonderful holiday; the feeling of holding your first child for the first time; the moment you heard that terrible news. . . . Some experiences echo so widely and strongly through our brains that the changes in the synapses they cause will never be undone.

Memories, then, are formed when brain cells become linked through firing together—*cells that fire together, wire together.* And once cells become embroidered together in this way—by our experiences causing them to fire in concert—they become like a secret brotherhood of remembrance. Solidarity is the essence of any brotherhood—one for all and all for one; strike my brother and you strike me. This Hebbian solidarity means that if my brother cell reacts, then I will react too. The brotherhood of cells develops a group mind: it just needs one to act to trigger all the others into action. It is because of this all-for-one, one-for-all welding of cellular souls that you could remember the stories you read by a few clues about the context that I fed you.

The secret of how to improve your memory should be becoming clear now. The bigger the brotherhood of cells that the memory is linked into, the more likely it is that you'll remember it later. One easy way to recruit millions of new members to this brotherhood is to link images in all the senses—sight, sound, touch, smell, taste—with whatever it is you have to remember, be it faces, facts, formulas, figures, or files. You want to make sure that you remember someone's face—some senior figure in your organization, for instance? Picture her face with a big floppy pink baby's bonnet around it, a drop of milk dribbling down her chin. You want to make sure that you remember her name—Susan Fell? Picture her toppling over a cliff.

You get the picture? The moment you picture the baby hat around her face and link her name to the image of her falling, you are triggering whole new networks of cells in the parts of the brain responsible for imagery. And as you know, much imagery happens in the brain areas where the real image would be seen. This means that, for instance, when you visualize Ms. Fell falling off the cliff, you switch on cells far back in the left half of your brain *as well as* cells in the language centers farther forward in the left hemisphere and the memory centers deep in the brain's inside. Because these are all switched on at the same time by your imaginings and memorizings, they become linked into a temporary brotherhood that is much more extended—and hence powerful—than the brotherhood that would have developed had you not used visual imagery.

In other words, you now have a much richer *context* for the memory of Ms. Fell and her name. And if you meet her six months later, you can impress her by remembering her name. Why? Because as you struggle to remember it, the fragments of context—". . . it was at that sales conference . . ."—will be linked by Hebbian principles to many other fragments of context, including the images that were stored with the original memory. Just the vague images of that sales-conference venue and memories of your mood and thoughts at the time will switch on the stored-up images of this woman falling and of her baby hat. And once you have these, the key thing to be remembered—her name—will drop sweetly into your consciousness. But it wouldn't have dropped in half so sweetly if you hadn't used your mind's eye to boost the brotherhood with which her name was linked.

Tricks of the Trade—Tips for Using the Mind's Eye to Boost Memory

Your memory will improve to the extent that you link it to different thoughts and sensations. Imagery in the mind's eye is one of the most ancient and effective ways of doing this. I'm not going to go into all the details about how to improve your memory—there are literally scores of books available that do that. I'm just going to give you some examples of techniques of imagery that boost memory if you use them.

Many of these methods were developed by the ancient Greeks—they named them *mnemonics*—and they used them to impress friends and senators by giving long and convincing speeches without reference to any notes. Given how important rhetoric was in classical Greece, this was a pretty powerful strategy for social advancement. One method the Greeks really liked was called the *method of loci*. Try this ancient memory method yourself now.

Attempt to remember the following list of words, but do it in the following way. Pick some path or route that you know well—from your house to the nearest store, for instance, or from a parking lot or station to your work. Do a mental walk along that route to make sure that you have the image relatively clearly in your mind. Now, take the following list of objects, and mentally lay each object at a particular point on the route—a corner, gate, storefront, etc. Mentally place each object at a different location. You might remember that the Russian memory virtuoso S that I described in Chapter 2 used this method. You can keep looking back at the list of words while you are placing the objects—you don't have to memorize them verbally first and then lay them out over the route.

Here are the words:

> spoon, glove, frog, cactus, wrench, mirror, cup, knife, hat, ring

Did you manage to lay these out on your mental street? Mentally walk with your mind's eye along this path again, and see if you can "see" the objects in their places. Check your memory for the list of words against the list written above.

This is a—literally—classic example of the power of the mind's eye in boosting memory. By visualizing the words along a familiar route, you bind the memory traces for these words into a much bigger brotherhood of brain cells than if you just learned them in the normal way.

Another very old method is called the *pegword mnemonic*. To use this mnemonic, you have to learn a nonsense rhyme by heart. In this rhyme, each of ten digits is linked with a rhyming object. Take a couple of minutes to memorize this:

One is a bun,
Two is a shoe,
Three is a tree,
Four is a door,
Five is a hive,
Six is sticks,
Seven is heaven,

Eight is a gate,
Nine is wine,
Ten is hen.

The rhymes make this easy to learn. Make sure you have it by heart before you read on.

You can now use the images linked to each number to remember lists of people, objects, or other numbers. Take this example. Imagine you're at a meeting and you have been introduced to ten people. How on earth do you remember all their names? It's not too difficult, if you follow these steps.

1. Picture the people around the table or room in your mind's eye.
2. Try to generate an image for each of the people in the meeting, linked to the number pegwords. I've given a few examples below, but these are completely arbitrary—you can use any images you like.
3. Go around them in order, starting on your left. Suppose that is John. Imagine John biting into a huge bun, crumbs dribbling down his suit. If you like, you can add a link to your mind's ear—a mother's voice scolding him with, "Oh, John, don't make such a mess."
4. Say the next on the left is Sheila. Use the two-shoe peg to picture Sheila sitting in a giant shoe. Again, you might imagine a voice asking, "Where's Sheila?"
5. Joe is third on the left, and three-tree can easily generate an image of Joe sitting in the branches of a tree, with someone shouting up at him, "Joe, come down!"
6. Peter is next, and four-door could see him, for instance, swinging on the door like a child, chanting, "Peter's swinging, Peter's swinging."

7. Celia in fifth place could be having trouble with a hive of bees buzzing around her, Fred in sixth place might be chopping sticks beside a roaring fire, Colin in eighth place might be hammering at a gate, while Jean in ninth place might be slumped beside an empty wine bottle in your mind's eye. Chris in tenth place might be sitting demurely with a hen roosting on his head.

You get the idea, no doubt. Though there is no need for bizarre or silly images, anything that you find funny or interesting will be better remembered because the faint ripple of emotion generated in your brain will recruit a few thousand more brain cells into the brotherhood of that particular memory.

Now try your own images, this time to remember the people around a table at dinner—they are all strangers, and you are struggling to remember their names, which are:

> Jill, Mark, Mary, Lorna, William, James, Felicity, Fiona, Mike, Steve

The pegword mnemonic is also very useful for learning numbers such as telephone numbers, passwords, or PIN numbers for bank or credit cards. You use them in exactly the same way as for learning lists of names or objects. Take the number 1982345. You can learn that by creating the image of all the objects related to these numbers in a row. Namely: bun, wine, gate, shoe, tree, door, hive.

At first this will take longer than learning the numbers in the usual way. But with practice, you'll get much better at using your mind's eye in this way, and not only will you remember numbers more quickly, but you'll be much less likely to forget them. These are examples of how you can use the power of the mind's eye to boost memory. There are whole programs that use the same principle of linking memories to visual and other images, which you can read about if you are interested.

It's really quite astonishing that we don't teach our children to use these mental-imagery tools as part of the curriculum. They could be as useful as reading, writing, and arithmetic in building mental abilities that will boost memory, learning, thinking, and creativity throughout their lives. And they are also critically important for managing that other affliction of Western societies—stress, the topic of the next chapter.

7

Vistas of Stress

At the sight of the speaker's platform, his heart began to beat a little faster. Looking down, he saw his notes, crumpled and damp from the sweat of his tightly clenched hands. His stomach tightened, and he had to take some deep breaths to try to calm himself. Boom . . . boom—*the heartbeat audible even to others, it seemed to him. The speaker was showing her last slide and the chairman was looking down at his program, looking over and giving a curt nod. His stomach tightened another notch and the heartbeat began to spiral out of control*—boom, boom, boom.

He glanced around, then wished he hadn't. The sea of faces made him dizzy—they were all looking at him. The chairman was thanking the speaker for her excellent lecture, a burst of applause from four hundred people. His hands were tugging at his notes, frantically trying to smooth out the crushed, damp bundle. His eyes darted down for the first few words that he had rehearsed and rehearsed but were now lost in the black hole of a blanked mind. His gorge rose in a sickening wave of panic—his sweat had smudged the ink and the text was blotched and illegible. Oh God, oh God . . . Boom-boomboom—*the rushing thud of a heart out of control. It would surely burst on him.*

He looked wildly up when he heard his name echoing around the hall. The chairman was looking at him. The last speaker was staring down from her chair on the podium. He felt the gaze of the crowd on his neck. His legs were jelly—he couldn't stand up—he would crawl for the door, escape his only chance. But his name was being called again. Every single person's full attention was on him. His breath could only come in panicky, airless, panting gasps. Again they were calling him.

His shirt, sodden with sweat, clung to him as he stood up on shaky legs. He shouldn't have looked down again, for two great dark semicircles of sweat spread from his armpits right out to the breast of his linen jacket. He stumbled, and someone gave a little gasp—a half-stifled giggle somewhere else behind him. He made it up the stairs to the platform. He saw the chairman waiting for him, frowning, holding out the microphone, his fingers drumming on the lectern. He tried to walk toward him, but his legs were lead and jelly, paralyzed, holding him there.

An impatient rustling from the audience—whispered comments about him throughout the room. His notes no more than a sweaty mush in his hands. Moving as if through molasses, past the last speaker—all smooth and smug and unsweaty—his name being called again, the applause when it came sardonic. Gasping for breath like a netted fish, he swayed as his eyes lifted to the crowd.

Silence fell. All staring at him, expectant—some were smiling. Like the audience at the Colosseum sensing blood, sensing disaster. A chuckle as he dropped the ball of wet paper onto the table. His knuckles white on the lectern. He had to speak, to say something—but no sound came out. His lips moved, but only a hoarse moan came out, piped and magnified through the hundred speakers of the hall. And with eight hundred eyes upon him, he felt the hot tears of humiliation begin to roll down his cheeks as embarrassed coughs and stifled laughter swept through the hall.

He opened his eyes and reached for the phone.

"Is that Peter?"

"Yes, hi, Chris—all set for next week?"

"Look . . . Look, Peter, I'm afraid I'm not going to be able to speak at the conference next week—I'm afraid something's come up."

"WHAT?"

The Incubating Image

Have you ever avoided something because of what your imagination conjured up by way of disastrous consequences? Chris's career took a blow because he ducked out of the conference. And the reason he couldn't go ahead was because of how vividly he visualized the humiliation and blind panic of succumbing so publicly to raw fear. It's pretty common to be very anxious about public speaking, and if that's the case for you, then you'll very likely have felt a tremor of anxiety triggered by imagining the scene above.

Most people have their secret fear, their incipient phobia. What is yours? Heights? Enclosed spaces? Snakes? Needles? Blood or injury? Spiders? Mice? Worms? Vomit? Water? In my work as a practicing clinical psychologist, I have seen people with phobias in all these areas and many more.

Don't do this exercise if you have anything more than a mild problem with this fear. If you have had to seek treatment for it or if it is a major problem in your life, it's best not to do this. If, however, as in my case, this is an anxiety that does not interfere with your life, then take a moment to visualize and imagine yourself into a scene involving what you are frightened of. In other words, do what Chris was doing when he imagined his public-speaking disaster.

Try to imagine not just what you see—visualize what you hear and feel too. Imagine too what's going on in your mind and body—visualize the emotional currents running through you, just as Chris was doing. If, for instance, it is heights that tend to alarm you, visualize yourself standing at the very edge of a high building, or clinging to a rock face. Feel yourself into the vertigo—imagine the swaying sensation and the fear in response to this instability. Run through the thoughts, images, and feelings in your mind and body.

I've just done this, imagining myself—as I did incidentally in a dream the other night—locked in a tiny cell not big enough to stand up or stretch my legs in. As I imagined it, I felt my heart begin to pound and my palms became sweaty. I could feel—like Chris—the panic rising in my chest, and my breathing became fast and shallow. How about you? Did you manage to switch on your body and brain's emergency escape systems as I did?

Anyone who uses the mind's eye to visualize frightening scenes will re-create the way body and brain will react to the real thing. In most cases, it will be a pale shadow of the real thing, though for a minority like Chris, the visualized scene can end up being *worse* than reality. How can this be? Well, in the vast majority of events in the real world, the worst will *not* happen. In the world of images and imagination, however, the fear-stricken visualizer can all too easily create a horribly vivid enactment of the ultimate disaster.

And it's not just all in the mind. When researchers study people who—like me and some of you—have a mild phobia that doesn't need treatment and isn't interfering with their lives, they have shown how the body's defense systems quiver in response to the feared image. In one study, for instance, student nurses who were relatively fearful of snakes had to visualize snakes while wired up to devices measuring their heart rate. The hearts of both the fearful and the less fearful nurses beat faster when they imagined the snakes, and both reported a tremor of apprehension. But as they were asked to repeatedly imagine different snakes, the two groups diverged. Over time, the nonfearful group's hearts gradually stopped responding with a faster beat to the imagined image. The fearful group's hearts, on the other hand, grew more and more nervy with repeated imaginings. Toward the end, their pulses raced much faster as they imagined the snake than at the beginning.

As the mind's eye conjured up the feared image, it seemed to worsen the fear and incubate the phobia among the most fearful, like some malignant mental virus. This is exactly what had been happening to Chris. By repeatedly imagining the public-speaking scene with its disastrous consequences, he was incubating the fear by imagery, up to the point where he couldn't go on with it in reality, to the detriment of his career.

And such vivid imaginings change your brain activity dramatically too. Six bank workers who had recently witnessed a bank robbery were studied in a PET brain scanner. The researchers showed them a video of an armed bank robbery and also a video of a neutral event that didn't have the same emotional connotations. Compared with when they were watching this neutral video, their brain activity in response to the bank-robbery video was much greater, particularly in the visual areas toward the back of the brain. The visual vividness of the frightening scene was cranked up by its emotional connotations.

Malignant Pictures

Chances are that a few of these bank workers would have gone on to suffer long-term stress difficulties if they weren't treated. In any traumatic event, disaster, or life-threatening accident that causes immense stress to those involved, only some of the victims will go on to suffer long-term stress and associated health problems. For severe events, this figure can reach up to 50 percent.

One really well-studied event was the nuclear power station accident at Three Mile Island in Pennsylvania in March 1979. Because of human error and mechanical failure, the reactor core was seriously damaged, and radioactive gas was released. The accident left a time bomb of a vast building filled with radioactive gas and half a million gallons of radioactive water.

In an atmosphere of confusion and steadily growing panic, news leaked out over several days. Only at the end of the week were pregnant women and mothers with young children told by radio stations to get out of the area. Government agencies and the company were meanwhile issuing often contradictory statements, and in an atmosphere of chaos and lack of trust, people began to doubt that things were under control.

After two weeks, the evacuees were told they could go back home, but they did so amid debates and disagreements about decontamination and what to do with the trapped gas. Meanwhile, the image of the looming, brooding threat of the nuclear plant's cooling towers was featured in every TV broadcast and on every newspaper's front page, an icon that etched itself into the vulnerable brains of the stressed and frightened residents.

On top of that came all the debates among experts about the cancer and other health risks that might already have occurred in the immediate aftermath of the accident. These worries were not helped when, sixteen months after the accident, the decontamination problem was "solved" by releasing the radioactive gas into the atmosphere in what were euphemistically described as "controlled bursts."

Fifty-four of the hapless residents living within five miles of the power plant were studied for many years after the accident. They were compared with another group from a low-risk area eighty miles away. The findings were dramatic—even eleven years after the accident, the traumatized group was still suffering significant stress compared with the control group. This was not just stress based on their subjective reports; it also showed up in stress-related chemicals in their urine and poor concentration on tests of attention. Only roughly half of the group who lived near the plant succumbed to prolonged and abnormal stress. What distinguished the people who succumbed to stress from those who didn't?

The first answer was that the people who felt helpless and out of control of events fell prey to stress. The second was that the people who experienced intrusive images and memories of the disaster were the ones vulnerable to chronic stress and all its associated effects on health.

Five years after the accident, many of the Three Mile Island residents were still haunted by images of the disaster, as well as memories of the panic and worry they felt at the time. It was as if the nuclear power plant was exploding again and again—only in the mind's eye rather than in reality. But as you've seen in this chapter, mental images conjure up much the same fluster of brain and body activity as do real events. In fact, as far as these stressed-out organs are concerned, in many respects the nuclear power station might as well be exploding over and over for all the difference it makes to them.

Five years later there were still many individuals who—like Chris at the beginning of this chapter—were haunted by reliving in the mind's eye what they feared. And each time the memory came into mind, their hearts would beat faster, their palms would sweat, and the noxious chemistry of chronic stress would sour brain and body. A further eighteen months later—no less than six and half years after the nuclear accident—those individuals who had still been tormented by visualizing the accident eighteen months earlier were much more stressed than their neighbors who had managed to ward off these images of fear.

Many soldiers who endure active combat develop post-traumatic stress disorder—chronic anxiety with "flashbacks" to the trauma and many other disabling psychological and physiological problems. Left untreated, PTSD can result in violence, severe depression, marital breakdown, and—in a minority of cases—suicide.

When combat veterans who have PTSD are exposed to reminders of combat—battle sounds, explosions, shouts—it triggers vivid imagery of the terrifying scenes. And there are a flood of changes in the body that don't happen in combat veterans who escaped PTSD. The traumatized veterans break out in a sweat at these sounds, their heart rate rockets, and fight-or-flight brain chemicals like noradrenaline pump into their bloodstream and brain. In other words, they react to the mental images the way you would react if your car went into a spin on an icy road and you just managed to avoid colliding with a truck.

What's more, even when veterans were not visualizing these combat scenes, their brain and body chemistry was chronically soured with a bath of stress-related substances such as cortisol. Cortisol is pumped out when you are under stress, and if the stress is severe and goes on for many years, this substance can corrode connections in the brain's memory centers. Studies have shown that these parts of the brain—known as the hippocampus—are shrunken in sufferers of severe and chronic PTSD.

Even almost twenty years after the battles in which these young men fought, many of them are still sick and traumatized from the experience. When researchers read scripts to them of battle scenes based on their own actual experiences, their brains and bodies react in the dramatic ways just described. Here is an example of a script, taken from a young man about a battle in 1968 and read to him in 1986, eighteen years later:

> It's May 1968. You are attacked just south of... by a divisional-size force. The sky lights up and the enemy attack using hand grenades. Hand-to-hand combat breaks out around you and you feel so frightened that your teeth can't stop chattering. Something hits you on the head—a grenade—your heart begins to pound. You grab it and throw it, but it explodes and you are wounded. Your whole body shakes.

But not all Vietnam veterans who endured these typical battle experiences developed post-traumatic stress disorder. Ex-soldiers who were not sick and disabled by this disorder, but who had been through equally severe trauma, were also read personalized scripts about particular battle scenes they had told the interviewers about. While they could picture them pretty vividly in their mind's eye, their bodies did not react with the violent symptoms of fear that the PTSD soldiers did. It was as if, the researchers reported, these veterans could remember without reliving the experience. They could recall with a relatively calm detachment the particular scenes, but they did not become *absorbed* in the scenes to the point of reliving them along with all the bodily symptoms. In what seems to have been a two-decade-long haunting by these flashbacks, the soldiers with PTSD endured severe psychological problems, physical ill health, and huge relationship problems. Why some young men succumb in this way to the trauma of battle, while others do not, isn't yet clear. But it may have something to do with the imaginative power of their mind's eye to allow them to become repeatedly reabsorbed into these hellish experiences.

One post-Vietnam war study showed just that. Researchers in rural Utah advertised in the local media for Vietnam veterans. Of the twenty-six who volunteered, fourteen had high levels of post-traumatic stress—even though their experiences in Vietnam were almost a decade old by this time. The researchers also measured how vividly these men could create images in the mind's eye—not of battle or trauma, but of neutral, irrelevant scenes, using a questionnaire very like the one you filled out in Chapter 4.

The results were remarkable. The traumatized men who were still haunted by their experiences had a far higher capacity for mental imagery than their comrades who had been through equally harrowing experiences but came out of them much less traumatized. There are only two explanations for this: either post-traumatic stress boosts the power of the mind's eye and enhances mental imagery for images that have nothing to do with the trauma; or the blessing of powerful mental imagery can turn to a curse when it comes to reliving trauma.

The latter is a much more plausible explanation. If anything, stress tends to reduce mental capacity rather than enhance it. It's much more likely that a lively mind's eye can all too vividly burn into the mind awful images of a trauma witnessed. And once stored, these pictures, it seems, can perpetuate a hellish and recurrent reliving of these scenes in such a way as to prolong the anguish and stress. Of course this is not the only reason why some people succumb to post-traumatic stress and others do not, but it seems very likely that—as with all potent technologies—the machinery of the mind's eye can inadvertently be harnessed for harm as well as healing.

"I Have to Worry So It Doesn't Happen"

Are you a worrier? If someone from your family is late home, does your mind go into overdrive, concocting all the possible disasters that might have happened? Ask yourself this: Behind it all do you think that worrying might actually ward off the disasters you are worrying about? If you're not a worrier, ask someone who is.

Whether or not you get a positive answer to the question, the fact is that worry may in fact help ward off disaster. Or, to be more accurate, worriers worry to fend off *images* of disaster that might otherwise be projected onto the screen of the mind's eye. This is because worry is primarily a language-based mental activity, where imagery is kept to a minimum. Worriers ruminate about catastrophes, but they do so in the abstract, via the relatively bloodless and imageless "cool web." We know that language can inhibit the projection of images in the mind's eye, and the word-based business of worrying seems to be no exception.

Close your eyes and relax for two minutes. Sit back on the chair where you are sitting, and let your mind wander. Occasionally, during this short time, note what's going through your mind.

What was in your mind when you checked? Thoughts? Images?

It is normal to have a mixture of thoughts and images going through your mind. But people who suffer from chronic anxiety tend to have very few images when they let their minds wander during such a short spell of relaxation. Rather, their minds are full of words and relatively imageless thoughts. Once they are successfully treated for their anxiety, however, they return to the normal pattern of half-word, half-image in the random junkyard of the wandering mind. Worriers worry precisely because it wards off—in the mind at least—the situations they fear. Worry neutralizes images.

The problem is, as we all know, that unless we face up to the things we fear, they will go on tormenting us. Worry is a way of avoiding facing up to mental images of what we fear. Because these images are facsimiles of the real thing, avoiding the images is like avoiding the reality.

Close your eyes for two minutes, and think about anything except sheep.

What did you think about for that two minutes? Did you manage to prevent the word *sheep* or images linked to sheep from coming into your mind? Most of you won't have managed to do this. This is because trying mentally to avoid something requires you to think about that something in order to avoid it. Telling you to think of anything but sheep immediately switches on all the brain networks into which the words and images linked to sheep are woven. And once these networks are active, they will hurl these into your consciousness, in spite of your best efforts. Paradoxically, then, you end up being *more* likely to think sheep-thoughts when you are trying not to think them.

The same is true for the things that torment us with anxiety. If we try to push them out of mind, they will spring back with a vengeance. This may be why worriers are worriers: If we crank up brain systems that inhibit imagery—namely word-based thought—then this will temporarily switch off or at least tune down the fearful pictures in the mind's eye. And sure enough, if you get people who are nervous about public speaking to worry for five minutes, and then to try to picture in their mind's eye a frightening image of public speaking like the one Chris imagined, then their bodies do not react to these images with pounding heart. In contrast, have them think neutral thoughts for a few minutes and then visualize the frightening scene, and their hearts will pound and their palms sweat when they picture the scene.

The only real way to overcome such fears is gradually to face up to them—both in the mind's eye and in reality. More of this later in the chapter. Worrying is a bit like a short-acting tranquilizer—it wards off anxiety as long as you use it, but isn't an answer in the long term.

Other Emotions in the Mind's Eye

Image and emotion are intimately linked, for good and ill. Language has a much more distant relationship to emotion, and so if we want to control our emotions, we have to access them through images. Beating stress and other unpleasant emotions depends on learning to use the mind's eye. This can be hard for some people who haven't cultivated this ability, but as you have seen earlier in the book, you can train yourself to be a better imager. Vivid images can perpetuate bad emotional states, but they can also help beat them. Let's turn to some other emotions. Have you ever felt guilty? If so, try this exercise. However, don't do this if you feel depressed, or you have been treated for depression.

Everyone has felt guilty at some time. Try to remember an event that you felt guilty about. Imagine yourself actually doing what it is that made you feel guilty. Close your eyes and try to reexperience that event.

Did you manage to re-create the situation that you felt guilty about? Did you feel any guilt this time? If you did, then your brain would have shown a distinct pattern of activity in the front part of the temporal lobes—at your temples on each side of your head. Low down on the surface of your left frontal lobe, tucked in above your left eye, there would also be a brain hot spot, and deep in the middle of the brain in a structure called the anterior cingulate, there would also be big guilt-driven activity.

In this last area, located near the emotional centers of the brain, the emotional turmoil stirred up by the guilt was being enacted. In the outside surface of the brain at your temples, you would have been reliving the events in question, dredging them reluctantly from the memory banks located in these storehouses. And in the left frontal lobe, your brain was frantically trying to deal with this temporary turmoil, trying to inhibit the most painful emotional spasms, and generally trying to regulate this storm of guilt. That, at least, is one possible story to explain the surge of brain activity that guilt evokes. When you imagine yourself in situations that produce different emotions, your brain and body react in quite distinct ways.

Imagine a time when you were really angry. Close your eyes and try to conjure up the images of the events that made you so angry. Once you have the picture in your mind, just pay attention to your pulse and other bodily sensations.

If you successfully imagined a scene that made you angry, then your heart would begin pumping faster and more vigorously, sending blood to the muscles that would dearly like to act out the anger.

Summoning up sad images into the mind's eye, however, actually turns down the output of your heart, and it also changes the state of your brain. These sudden downward tips of mood, which everyone can experience sometimes, often happen because some unbidden memory image erupts into awareness—maybe triggered by an event or reminder that you are not even conscious of.

In the laboratory, psychologists can easily change your mood. They can make you sad (with your permission, of course) by having you read a series of statements like this: "Looking back on my life, I wonder if I have accomplished anything really worthwhile. There are things about me that aren't very attractive. I feel a bit disappointed about the way things have turned out."

Once you have read through a few dozen of these sentences, your mood sinks for a few minutes. You can similarly tilt people into a happy, elated mood by getting them to read sunny sentences like these: "All in all, I'm pretty pleased with the way things are going. Life is pretty good at the moment. Everything looks good the way I feel right now."

Music can also help. In one research project, people who had read the gloomy sentences also had to listen to Prokofiev's brooding music *Russia Under the Mongolian Yoke*—played at half speed! Those who had read the cheerful sentences listened to Delibes's *Copelia*, or to an anodyne compilation of popular classics.

Perfectly normal volunteers can be tipped from elation to sadness and vice versa by these techniques. Of course, these mood changes are only temporary, but the changes in the brain that go hand-in-hand with them are very real. Brain activity changes significantly, particularly in the frontal lobes. In particular, it seems that sadness increases activity in the right frontal lobe, while happiness or euphoria increases activity in the left frontal lobe.

Visualizing disgusting images also produces brain and body changes, which include a pumping heart and a tensing up of the muscles—particularly around the lips. One scene that volunteers had to close their eyes and visualize in a study of disgusting imagery was this: "Feeling hungry deep in the woods with no food; my best friend reaches down, picks up a dead bird, and bites into its bloody body."

It is uncanny how vividly you can visualize and experience

a situation like this in the mind's eye, right down to the physiological responses that you would have if you really did witness such an event. It is the very potency of this capacity that makes the images projected on the mind's eye a force to be reckoned with in all our lives. And this can be a force for good and bad. Chris, the panicking public speaker at the beginning of this chapter, could probably have benefited from duller images in the mind's eye. Others can use it for more enjoyable purposes. Men, for instance, asked to imagine sexually arousing scenes, produced a very realistic and distinct pattern of brain and body activity. Interestingly, the pattern of brain activity was very similar to that shown when they visualized competing with other men in nonsexual activities!

Once Bitten, Twice Shy

If a dog bites a child, it is pretty common for that child to become frightened of dogs. This is because of a fundamental type of learning in the human brain that has been very useful in ensuring our survival as a species—conditioning. In all of our brains is a vast tangle of conditioned responses. You feel hungry when you open the fridge? This is because the stimulus of opening the fridge has in the past gone together with feeling hungry. Do that a few times, and the fridge opening will trigger brain and body hunger signals, even though your stomach may be full.

The same is true for smoking. Smokers often feel a particular craving for a cigarette after a meal. This again is because the end-of-meal stimuli—let's call them cues—have repeatedly been linked in the past with the sensation of smoking and all the physiological baggage it brings with it.

It's the same for pleasures such as sex. If sexual arousal is triggered along with particular cues, then these cues can gain the power to trigger sexual arousal. This is how fetishes develop. There is nothing inherently sexually arousing about shoes, but there are hundreds of thousands of shoe fetishists around the world. This was vividly demonstrated by one of my teachers, Professor Jack Rachman, at the Institute of Psychiatry in London, thirty years ago. In an experiment, he showed slides of nude women to a group of his male students, while at the same time showing them women's boots. The students, who had no particular interest in women's boots before this, gradually began to become aroused at the sight of just the boots, even when there were no nude women to go with them. The natural sexual arousal to the women had become conditioned to the previously innocuous sight of the boots. In other words, their professor had instilled an experimental fetish in his students. (I hasten to add that I was not one of the students in question.)

Human beings are not salivating Pavlovian robots, but in our brains there are literally thousands of automatic conditioned responses linking emotions and their concomitant physiological reactions to various cues inside and outside of our bodies. You feel a surge of anxiety when you go into a crowded room of strangers? Your stomach contracts when you hear the whine of a dentist's drill? You crave a drink when you arrive home tired after a hard day's work? These are all examples of conditioned responses, brain and body reactions that have become linked in your brain to particular cues and events.

And we are talking very powerful physiological reactions here—not just diffuse thoughts and feelings. A powerful example of this comes from work on heroin addiction. Heroin addicts gradually develop a tolerance to the drug the more they use it. In other words, just as a couple of drinks makes the teenage drinker tipsy but leaves his dad untouched, so the experienced heroin addict needs more and more heroin to produce the same effect on his body and mind that a small amount would achieve for the novice.

This happens in rats also. They can tolerate larger and larger amounts of heroin—quantities that would kill a novice addict. But this isn't a purely physiological change that happens without relation to the environment. The body's adaptation to the drug is conditioned to the signals and cues that are linked to receiving it. Specifically, if the rats are always given the heroin when they are in a particular box—say one with green walls—the tolerance to the heroin is *conditioned* to that box. Give the hardened, highly tolerant rats heroin in a new box—say one with red walls—and the dose can kill them.

A similar thing happens in human addicts. Many unfortunate addicts die of accidental overdoses after they have been off the drug—say in a rehabilitation clinic—for a while. They then go out—say to a new home away from their old haunts—and succumb to the temptation to take the drug again. But they may take a dose at a level they had become used to before going on the wagon, and—because there are no cues to trigger the body's protective tolerance reactions to the drug—it kills them.

During the Vietnam War, the U.S. government was very worried about the vast proportion of the U.S. armed forces there who were regularly using hard drugs such as heroin. An epidemic of heroin addiction was feared on their return. In fact, most of the men who were regular drug users in Vietnam didn't continue the habit back home. One reason for this was that the addiction to the drug—with all the associated craving and tolerance—was conditioned to a very particular set of cues: the sights, sounds, smells, boredom, and fear of Vietnam. These cues were not at all present back home in the States, and so the craving and habits linked to heroin were not present. No cues, no craving. No craving, no addiction.

Your Miniaddictions

This exercise will only work if you have at least a little bit of an addiction—whether that be for nicotine, alcohol, chocolate, gambling, sex, exercise, or some other substance or activity. Even if you are an addict, it may not work in the particular circumstances in which you are reading this book. I, for example, am partial to the odd drink of alcohol, but as I write this—at eight o'clock on a Friday morning—the last thing I feel like having is a pint of Guinness. If this were eight o'clock on a Friday evening, however, this exercise might work for me. Why? Because almost all our urges and cravings are conditioned to cues: I have never drunk in the morning, but have usually taken a drink on a Friday evening—so Friday evening is a strong cue linked to the effects that one or two pints of Guinness have on my brain and body.

Try this with your miniaddiction of choice. Of course, if you have had an addiction problem in the past for which you've had treatment or that has caused you problems in your life, it would be best not to tempt fate by trying this. Pick the circumstances appropriate to the urge for this imagery. I, for instance, will visualize myself coming home on the train on a Friday evening, tired and thirsty.

Close your eyes and visualize the scene most linked to the urge in question. A smoker might imagine herself drinking coffee after a good meal. A gambler might visualize himself at the racetrack, hearing the commentator's excited voice over the loudspeaker, seeing the arcane hand signals of the bookmakers. A binge eater might imagine himself in a bakery, surrounded by chocolate-and-cream cakes, then biting into one, feeling the clingy, yielding sweetness of rich chocolate on his tongue.

Whatever your weakness, try to re-create the sensations linked to it—the internal and external cues—in your mind's eye. Don't forget the other mental senses also—try to feel, taste, touch, smell, and hear as well.

Having tried this exercise last night, I had no difficulty imagining myself onto that Friday-night train, and could create the thirst at the back of my throat that was entirely absent before I did this exercise. I could even feel, vividly, my fingers curled around the cold glass.

Okay, so much for my confessions—how did *you* do? Could you re-create the sensations and bodily reactions linked to your miniaddiction? Most of you will have been able to do so, at least to some extent. And though this was an exercise of the mind's eye, the effects would in most of you have spread far into your body and brain. Just as having the chocolate cake there in front of your eyes makes your mouth water and your pupils dilate, so the mental image has the same physiological effects.

When the heroin addict sees the syringe, his brain chemistry changes, even before the drug comes near his bloodstream. The more hardened an addict he is—in other words, the more heroin he needs to achieve a hit—the bigger will be the size of his brain's anticipatory chemical lunge. This is called an *opponent process*, and is part of the body's survival gear. So that it is not too destabilized by the brain-altering drug, the brain prepares for the drug's arrival by acting out the physiological mirror image of the drug's effects. So, for instance, if the chocolate cake usually soothes you, the sight of it will create a slight anticipatory edginess—the opposite of soothing. When you actually eat the chocolate, the real effects will be partially canceled out by the opponent process—the mirror-image brain-body response that has been triggered by the cues linked to the chocolate. In the case of the heroin addict, the cues linked to the heroin—for instance, the sight of the melted heroin bubbling on the spoon—will cause a more dramatic edginess, a sort of miniwithdrawal symptom, as the brain and body pull back with a mirror-image response that will cancel out some of the real response caused when the drug hits the brain.

The more addicted you become—whether it be to nicotine, heroin, or chocolate—the smaller the difference between the real and the expected effects. In other words, if you begin to stuff yourself with chocolate every day, or take several high-punch fixes of heroin per day, the brain and body will steadily up-regulate their counterbalancing response when they expect the drug to appear. This happens to a point where there is little positive effect to the drug or activity, and the whole business becomes a vicious cycle of fighting off the opponent processes triggered by cues in and outside the body as a result of constantly taking the drug.

It is when the drug or activity becomes motivated by the desire to avoid the negative effects of not taking or doing it that you really know that addiction has kicked in. Many cigarette smokers smoke largely to avoid the nasty feelings that come from not smoking. Many alcoholics drink not because it gives them much pleasure but because it wards off the greater evil of withdrawal symptoms. Many binge eaters eat to ward off feelings of guilt and tension that build up when they are not eating.

Being rewarded by something nasty *not* happening—"If you do your homework, I won't make you clean up your room"—is a particularly powerful method of learning. Once the brain slips into this mode of taking the drug or engaging in the activity—to avoid pain more than to gain pleasure—the habit is very hard to undo. This is one reason why it is so hard for chronic addicts of all types—of smoking, heroin, and alcohol especially—to give it up. Another reason is that most of our mini—and maxi—addictions become such because they produce pleasurable effects on our brains and bodies, at least in the short term. So when the brain learns to produce the opposite effects in anticipation of the drugs, these opposite effects tend to be relatively unpleasurable: edginess rather than calmness, hunger rather than satiety, muscle tension rather than relaxation, etc.

The problem is that these negative symptoms are very like a lot of other rather general symptoms related to stress, low mood, tiredness, etc. And because they are very alike, they can often be mistaken by our brains for the similar symptoms of craving. In other words, even though the chocolate cake might not be in front of you, its rich smells not curling into your nostrils, if you are feeling tense and edgy for some other reason, your brain might mistake these cues for those of incipient chocolate. If so, then the whole carnival of brain and body symptoms associated with craving your weakness will be switched on by these initially irrelevant feelings of stress and tiredness.

What's happening here is that sets of cues *within* your body are being stitched together by a process of conditioning. If you tend to drink alcohol when you are tired or fed up, the physiology of tiredness and fed-upness becomes embroidered into the same tapestry of conditioned responses as the cues linked to drinking. And so a whole network of linked internal and external cues spreads like a stain across the tablecloth of your emotional life.

For the U.S. soldiers in Vietnam, it was quite easy to wash up the stain: They moved home and away from an environment of stress, boredom, and quite different sights, sounds, and smells from those of home. For the rest of us, however, whether our addictions be big or small, it can be very hard to contain the spreading stain if we let our addictions come to the point where we do what we do in order to stop feeling bad. Once that happens, we are in a completely different ball game from the situation where we do what we do in order to temporarily boost our mood.

And in line with this argument, if you ask cocaine addicts to visualize themselves under stress, this causes a big surge in their craving for cocaine, even though the stress images had nothing to do with the drug. For these addicts, a whole raft of bodily cues linked to stress have become triggers for craving and using cocaine. And because addicts usually end up leading stressful lives, this means that the craving is on a hair trigger and is switched on by these cues again and again throughout the day.

A lot more can happen in the mind's eye when it comes to stitching together these networks of conditioned cues and responses.

Pavlov's Eye

Have you been to the dentist recently? I have, and I didn't like it much. What I like least is the penetrating whine of the dentist's drill. There is, however, nothing special about that sound that makes it unpleasant. The sound triggers negative sensations and emotions in me because that sound has become the conditioned stimulus for pain and discomfort. That sound-emotion link is one of tens of thousands of conditioned responses that have been welded into my brain because the stimulus (drill sounds) and response (pain, discomfort, and occasional anxiety) have occurred at more or less the same time.

Imagine you are lying back in the dentist's chair. Smell the antiseptic smells and feel the wads of cotton packing your gum. Feel the white coat of the dentist brush against your hair as she bends over you, and watch her hand reach out to the drill and pull the gleaming metal down to your mouth. Her fingers lift your lip, and the drill whine begins. You clench your hands tight as the drill bites into your tooth—a jab of pain and the slight smell of burning in your nostrils. Your hands are gripping the chair arms tightly and your whole mind is focused on that single needle of pain in your mouth. A plastic tube is forced into your mouth, and there is a bubbling, sucking sound as it extracts the blood-tasting saliva from your mouth.

Apologies to the dental phobics among you. As I ran through this imagery exercise myself, my heart beat faster and I began to salivate. Even now, after I have opened my eyes, I feel a residual extra trepidation about my next visit to the dentist. It feels as if I could really build up my anxiety levels if I allowed myself to keep on visualizing scenes like this.

This is precisely what happens to many people who have fears and phobias. By repeatedly visualizing the things they are afraid of, they strengthen the links between stimulus and response in the brain. If I were to continually visualize the sights, sounds, tastes, and smells of the dentist in my mind's eye, I would reinforce the links between the cue—the whining sound of the drill, for instance—and the emotional/physiological reactions that occur in the dentist's chair. Would that matter? It certainly would. When I get to the dentist again next week—if in fact I don't do what Chris did and cancel—the sound of the drill and the sights and smells of the surgery will trigger much stronger reactions than if I had not rehearsed the dental experience many times in my mind's eye. I will have a far greater emotional and physiological reaction in brain and body because the dentist-surgery cues will now be much more strongly linked in my mind to pain, discomfort, and anxiety, all as a result of this mental rehearsal.

This is partly what happened to the Three Mile Island residents who incubated the stress and trauma of the nuclear disaster by frequently reliving it in their mind's eyes. This is what Chris, the public speaking phobic, also did. Reliving stress and fear in the mind's eye can be not simply as bad as experiencing the real thing, it can be worse. How could imagining what you fear be worse than coming face-to-face with it? Well, let's take Chris for example. Suppose he hadn't lost his nerve and had forced himself to speak at the conference. While it is just possible that the disastrous scenario he kept visualizing would have materialized, it is much more likely that, properly prepared, he would have presented a reasonable—maybe even an excellent—lecture. It's in the nature of phobia that the phobic person finds it very hard to think rationally about the feared situation. But what people visualize about the feared situation is almost always far, far worse than reality. As I'll explain later, the best therapy for phobias is to face up to what you fear several times. Facing up is much easier to do in reality than it is in imagination.

Let's assume that Chris had left the stage with applause ringing in his ears. Though he might well be anxious, there would also be positive emotions—relief, achievement, pride, etc. These would reduce his anxiety and produce a quite different physiological state in brain and body. And the cues linked to public speaking—sea of faces, lectern, notes, slide projector—would have their link with the conditioned response of fear and anxiety weakened as a result.

This breaking of links between cue and response is called *extinction*. There would be no better way of breaking the link between these public-speaking cues and his terrified response than if Chris could just bring himself to give a lecture a few times. The problem with Chris now is that, having backed out of giving the lecture, he can only continue to rehearse in his mind more and more scenes of public-speaking disaster and humiliation. And believe me, there can be no extinction as long as this happens in the mind's eye. His phobia will incubate like bacteria, the cue-response connections being ever strengthened as they are constantly followed by disaster on the unconstrained stage of Chris's mind.

This is particularly so because Chris needs to be able to give lectures. This means he is in a conflict—between his mental-image–fueled fear on one hand, and the feeling that he should be able to give lectures on the other. When you experience conflicting impulses like this, you become more anxious. When you are anxious, uncomfortable thoughts and images spring more easily into mind. These images and thoughts in turn add to the anxiety, and so on. That's the problem with doing what Chris did—avoiding what you fear. Paradoxically, you end up much more tormented by fear, because the thing you are frightened of will haunt the mind's eye. The images will generate distress and anxiety that may well be worse than the real thing.

Exactly the same thing would happen to me if I let anxiety or anticipation of pain prevent me from going to the dentist. Again there would be a conflict—this time between wanting to stop my teeth from rotting, and wanting to stop the pain associated with that whining drill. The images of the dental procedures would incubate in my mind's eye, triggering an anxiety that would become steadily greater the less chance it had to be checked by the reality of a dentist's surgery, where in fact most of the time there is no pain or discomfort. Most of the discomfort is anticipation, and anticipation consists of a set of mental images, cues that trigger conditioned responses of discomfort and anxiety.

Researchers have shown this type of incubation in action in the laboratory. Students watched a screen as various shapes were shown. Whenever a triangle appeared, a startling blast of loud, frightening sound rasped out from some speakers behind them. After a few such pairings, the triangle and the frightened, startled response of brain and body became linked in their brains. Now, it just took the appearance of the triangle for the students to react with a burst of startled fear and sweating palms.

The students were now split into three imagery groups, where they rehearsed three different scenes in their mind's eye. The first were asked to visualize a cat meowing and their reactions to it. The second were asked to visualize the horrible image of someone trying to stick a pin in their eye and their reactions to it. Finally, the third group were asked to visualize the triangle, the blast of sound linked to it, and, again, their reactions to it. In the last stage of the experiment, the volunteers were all presented just with the original shapes, sometimes with the blast of sound, sometimes without. The third group—those who had rehearsed, or incubated, the triangle-sound-reaction image in the mind's eye—showed a much bigger fear response, as measured by the sweatiness of their skin, than the other two groups. Even the group who had visualized the anxiety-provoking scene of someone trying to stick a pin in their eye didn't incubate the triangle-blast-fear response. But the group who again and again ran through this conditioned response strengthened these links in their brains. The result was that the artificial "phobia"—the triangle that was linked to the startling sound—became worse.

This is precisely what can happen to real phobias. Just a few nights before I was writing this section, I went to a concert of learner musicians at a music school, all adults. One woman got up, started to play the piano—quite beautifully, it must be said—faltered, and stopped. We all assumed she was going to restart. Instead, she stood up, saying, "No, I can't do it—I knew this was going to happen," and fled. No amount of cajoling would bring her back to the small and very sympathetic audience, who had just sat through some excruciating performances and were desperate for quality playing of the type she had begun to show.

I have no doubt that she had rehearsed this entirely imaginary scenario in her mind for quite some time before her flight.

So vivid had this rehearsal been that she fled a nonexistent disaster. She had incubated the conditioned response between disastrous performance and resulting fear (equivalent to the blast of sound for the students and the resulting fear) on the one hand, and the stimulus of playing piano in front of a group of people (equivalent to the triangle on the computer screen) on the other.

The students strengthened this connection by imagining it in their mind's eye, so that when they again saw the triangle, their conditioned fear was even stronger than before. Not only that, but when both the triangle and the sound appeared, they jumped out of their skin even further than before they went through the imagery. And so it was for the poor pianist. She had rehearsed the connections in her mind so often before the event that when the real stimulus appeared—sitting at the piano with an audience—her reaction was even worse than it was before she started mulling over these images in her mind's eye.

Conditioned responses can even be made in the mind's eye. A terrifying nightmare of a particularly horrible event happening in a certain situation can lead to that situation becoming a source of anxiety. In this case, a purely mental image becomes the cue conditioned to a very real anxiety, even though the scene in question never took place. I know someone who developed a fear of heights that emerged out of the blue, when she saw her friend standing at a parapet above a very high drop. Suddenly, she felt a wave of vertigo, and during this wave, a vivid image came into her mind's eye of her friend toppling over into the abyss. Her friend, perfectly safe, was perplexed to be pulled away from the edge. He was even more bemused to find himself begged not to go near any remotely high location after that. All of her friends and family became subject to the same entreaties. This person had developed a conditioned fear of heights because of a particularly vivid daydream-type image that came into her mind.

It is not just anxiety that can be conditioned in the mind's eye. As I showed you earlier, sexual desire can be conditioned too. Tragically, a small number of people's desire becomes perversely attached to sexual violence, or to children. Because these are criminal acts, in general these individuals get thankfully rare opportunity to act out their desire. This does mean, however, that they tend to act out their fantasies in vivid detail in their mind's eye, strengthening the links between the desire and the awful cues to which it has become conditioned.

Addicts—whether mini or maxi—can also incubate their particular cravings by mental images. This may be one reason why heavy-handed, obsessive dieting is so often unsuccessful. If you dramatically starve yourself—say by eating nothing but grapefruit for several days—you will inevitably stoke up images in your mind's eye of the forbidden fatty, chocolaty goo that is your downfall. And once these images invade your mental screen, they will incubate an even stronger link to the synthetic hunger that they have become conditioned to evoke.

Habits and conditioned responses can be incubated in the mind's eye, then. This is particularly true for those of us who have a vivid imagery ability.

Do you easily become absorbed in a book, film, or fantasy? Can you "feel yourself" into the situation you are witnessing or imagining? Take the dentist exercise earlier. Were you able to really be there in the dentist's chair, feeling, hearing, and smelling what was going on? What about books and films—do you enter into the plot so that you feel, hear, and see what the characters are experiencing, whether it be the sensation of the clinging mud of the swamp into which you are sinking or the feeling of a lover's hands on your skin?

The stronger your imagery ability, the more vulnerable you will be to incubating your fears in your mind's eye—and, for that matter, your sexual fantasies, your miniaddictions, and your aversions. The conditioned responses wired by learning into your brain will be much more readily strengthened—and weakened—through the drama played out in your mind's eye. This has its upside and its downside. The upside is that you might be able to beat some of the conditioned fears and cravings that you have with the help of your vibrant mind's eye. The downside is that you are more vulnerable to the incubation of these phobias and urges.

Research has shown that people who most often have images of their phobia are the ones who are most likely to avoid the situation in real life. If you are a snake phobic living in a big city, this shouldn't present too many problems. But if you are an insect phobic, or an agoraphobic, then all this avoiding can wreck your life.

That being said, I once had a patient—let's call her June—who was a snake phobic, and her life really was wrecked by avoidance. She had gone to work in Africa for a time, and one morning discovered a small snake curled up in her sandal, which she was just about to slip her foot into. This understandably traumatic event never happened again, but she became paralyzed by the fear of snakes, and soon left Africa to go back home to Ireland. As you may know, Saint Patrick rather efficiently banished the snakes of Ireland to the sea and so snake phobia shouldn't be a problem on that island. How was it, then, that June ended up unable to work, housebound, and on heavy medication for severe anxiety—all because of a persisting snake phobia? How could such a phobia affect her in a country where there are no snakes? Surely there was nothing to avoid and so no difficulty in living a normal life?

June's problem was one of a too-active mind's eye and an overdeveloped capacity for absorption in her imaginings. Logic has little power in the face of panic and severe anxiety, and this woman could not shake off the fear that she might find a snake under her bedclothes, behind the wardrobe, in her shoe, or in a drawer. She only had to catch a glimpse out of the corner of her eye of the coiled length of an electric cable for her to scream and run from the room. The more this happened, the more her mind's eye threw up images such as a cobra coiled among her tights in a drawer, a viper crawling up the bathtub drain, or an adder dangling from the curtain rod. Haunted by this torturing carnival of images, she ended up completely disabled by anxiety.

With therapy, June ended up resuming a normal life. She still was not fond of snakes, but she was no longer paralyzed by these terrible, irrational images. How did she beat this problem? Let's have a look at therapy in the mind's eye.

Therapy in the Mind's Eye

June's therapy was quite simple, and involved gradually exposing her first to images of snakes and finally to the real thing. We started off with pieces of string, ropes, and cords. At first, even the sight of a piece of cord on the table provoked the most intense anxiety in June. If she had been at home, she would have fled from the sight of the cord, but as she was in my office, I encouraged her to stay. Gradually, her anxiety declined to neutrality. After ten minutes, boredom had replaced it.

Next she had to touch the cord. This revived her anxiety—a jagged burst of fear, which, however, quickly subsided as she got used to the feel of the harmless cord. We then moved on to objects that were slightly more snakelike—pieces of hose, rubber-coated cable, and the like. Each time we had a new item, her anxiety rose but then gradually fell as she got used to it.

The difference here was that she was doing what she never did in her mind's eye—stay with the cue long enough to see that it didn't actually change into a wriggling, biting snake. The conditioned link between cue (long, potentially wriggly objects) and response (paralyzing fear) was gradually extinguished by presenting the cues while anxiety declined into boredom. Gradually the links that had been etched into her brain were weakened.

Over a few days, she progressed from tubes to cartoon pictures of snakes to pictures of real snakes. Then came a real challenge—having her first look at, then handle, a very realistic rubber snake. Near the end of her treatment, with trembling hands she picked it up, heart racing and soaked in sweat. After twenty minutes, she was calm and unconcerned, toying with the uncannily realistic beast. And at the very end of her treatment, a couple of visits to the reptile house at the zoo managed finally to break the link of fear that had been forged in Africa.

Therapy for Chris would follow similar principles—gradual exposure to cues linked to the scene, so as to break the cue-fear links in the brain. But what has this all got to do with imagery and the mind's eye? Well, if connections can be strengthened in the mind's eye, then they can also be weakened. In other words, therapy in the mind's eye is possible. Take this study of people with obsessive-compulsive disorder. People with this problem find themselves compelled to engage in repeated rituals. For instance, they may obsessively check again and again whether they have locked doors and switched off stoves before they can leave the house. This can be to the extent that it takes them minutes—even hours in extreme cases—to leave the house. Other people might be obsessively frightened of contamination after being in a bathroom and may have to clean their hands again and again, over and over, even to the point where they are red and raw from the repeated washing.

The most effective treatment for most types of obsessive-compulsive disorder is similar to that for phobias: You expose the person gradually to the cues that trigger the anxiety and the consequent rituals. You make sure that these cues are present as the anxiety falls, so that the cue-fear link in the brain is weakened. So, for instance, you have the person go into the bathroom, touch nothing, and do no washing of the hands. Then you might have them just touch the edge of the bath with a finger and resist any hand washing . . . and so on. We know that this type of treatment results in big changes in activity in the brain regions known to be overactive in obsessive-compulsive disorder, evidence that brain connections are being altered by this type of therapy.

But what about doing the same thing in the mind's eye? Why not ask these individuals to mentally expose themselves to the cues linked to their rituals, and then stop themselves in the mind's eye from going through with the washing, checking, etc.? One study found that mental training of this type produced longer-lasting effects than did real-life training for obsessive-compulsive disorder. These researchers argued that because so much of the obsessive habit took place in the mind's eye, the primary focus of treatment—at least for this particular group—had to be that mental screen.

Think of your miniaddiction again. Close your eyes and imagine it just within reach. If you are a smoker, feel the cigarette packet in your hand, pull the cigarette out, and hold it between your fingers. If chocolate cake is your weakness, imagine a wedge of it in front of you, the chocolate moist and yielding. Whatever your miniaddiction, make an image of it in your mind's eye, but stop yourself at the last minute from consuming it.

Did you manage that? Did you feel the urge to consume, and did you have to resist it? Close your eyes again and keep it up until you find the urge waning. It might help to use the power of your mind's eye to help you along by visualizing the cake turning stale. You might like to help the smoking urge by imagining a hoarseness in your throat at the touch of the smoke and visualizing the soft cells of your throat beginning sinister, precancerous changes.

Some of you will have managed this better than others. The vivid visualizers among you—those of you with the capacity for imaginative absorption—will have had an easier time with this exercise. The good news for those of you in this category is that you are better able to use the mind's eye to beat your fears and miniaddictions. The bad news, on the other hand, is that you are probably more vulnerable to incubating such links in your mind's eye.

Most of us have some ability to mirror reality in our mind's eye, particularly where the urges and fears close to our hearts are concerned. As I said earlier, I am somewhat claustrophobic. If I close my eyes and visualize myself locked in a small cell even for a moment, I feel my heart begin to beat faster and my palms begin to sweat. For most of us there is some situation that we fear, and for many of us there are urges we would rather control, whether for chocolate cake or something more exotic. We can use the power of the mind's eye to help master these unwanted fears and urges, if we want. The better you are at absorbing yourself into the multisensory virtual reality of the mind's eye, the better your chances of doing so.

Later in the book, I'll explain how most world-class athletes use the mind's eye for mental training, and this training is critical for actual physical performance. But it's not just athletes who can practice and boost their skills. Chris, for instance, could benefit from practicing public speaking in his mind's eye. He could start by imagining himself, for instance, speaking to an empty room. Then he could visualize himself giving his lecture to just one sympathetic friend. He could mentally walk the path from chair to lectern, visualizing the auditorium, feeling the notes in his hands, breathing slowly to calm himself, and generally enacting the feared scene on his mental screen.

He could gradually build up his mental-training regime until he was in the packed conference hall, feeling the sea of faces behind his head and noticing the chair's curt nod in his direction. He should mentally walk every step of the way, skipping nothing—right down to the feel of his best suit and the sweatiness in his palms as his hands lightly hold his notes. It may sound silly to some of you. "All in the mind," you might think. Yes, it is—but the fact is that when we imagine ourselves doing or feeling things "all in the mind," it has an incredible similarity to what happens in our brains when we actually do or feel these things. Indeed, as you saw earlier in the book, modern brain-imaging studies show that *in order* to imagine in the mind's eye, we have to use most of the brain machinery that would be used to do in real life what we are imagining.

So when Chris mentally walks toward the podium and imagines himself climbing the steps, the time it takes him to imagine this in all its detail will correspond closely to the time it would take him to do it in reality. Similarly, when he visualizes the lectern looming before him, the same visual areas in his brain are active as would be firing were he actually looking at that lectern. Imagery, then, re-creates an incredible facsimile of the world, and of your mind and body state within that imagined world. This makes it an incredibly useful and powerful forum for learning and habit-change. Try it.

You can even use imagery to combat the effects of electric shock—a useful technique, perhaps, for trainee spies and agents. If you get people to repeatedly imagine receiving a non-fatal but nevertheless painful electric shock, you can then test how easily they habituate to later real electric shocks. You can measure distress generated by the shock by changes in skin-moistness. When this is done, you find that the people who mentally inoculated themselves by visualizing receiving the shock show a much more rapid decline in distress as they endure real shocks.

Anxious about your next visit to the dentist? Or for some other medical procedure? Visualize yourself in the chair. Hear the whine of the drill, the smell of the burning enamel, the painful sharpness biting into your tooth. Visualize yourself relaxing rather than tensing up. Breathe slowly and calmly. Imagine yourself floating and detached while the drill whines. Visualize the pain as being somewhere disconnected from yourself.

It might not work the first time. Keep on visualizing this—take a break and come back to it if it is not working. Just keep on doing it until the image becomes neutral, even boring, rather than distressing. You are inoculating yourself, just as the people did who prepared themselves for electric shocks by imagining them.

This may not work for you, but it might—it depends on how vividly you can use your mind's eye, and how well you can control what appears on the mental screen. One critical feature of this is whether you focus attention on the outside or the inside world. When you attend more to your reactions—the sensations of fear inside your body and your *responses* to the scene—your body reacts with a greater burst of anxiety than if you focus your visualizing on the scene in the outside world. One problem for Chris was that he focused too much attention on his own body—the trembling knees, sweating hands, and pumping heart. Noticing these when you are anxious serves only to increase the anxiety. When you are inoculating yourself against the situations you fear, therefore, you should try to focus your visualizing attention on the scene itself. Any attention you give to your body should be imagining it calm and relaxed.

Millions of people throughout the Western world use psychotherapy to help combat their emotional problems. But much of this psychotherapy emphasizes words and not images. For many emotional problems, you need imagery just as much as words to overcome difficulties. As with the education system, our mental health industry has neglected imagery. It seems that this is true for our physical healthcare systems also: Let's turn to this in the next chapter.

8

Healing Pictures

Sheila had been plagued with painful genital herpes for ten years; it had played havoc with her life. The antiviral medication sometimes kept it at bay, but ten years is a long time to be taking drugs. If she stopped taking the antiviral medication, the herpes came back, again and again. If she was stressed or depressed, that was when it really came back with a vengeance. Her doctors told her that this was because the virus-fighting antibodies in her blood—lymphocytes and natural killer cells—hated stress and loathed depression; they would go on strike if they detected these emotions in her blood.

Then someone came along and trained her to create images in her mind of her body fighting the herpes virus. She was trained to relax herself under self-hypnosis and to visualize sharks and dolphins cruising through her bloodstream, snapping up the viruses. After a few weeks, the doctors measured her immune system: It had gone back to work—natural killer cells and the other virus-fighting predators were cruising through her bloodstream, scattering the shoals of herpes virus from her body. And the herpes didn't come back nearly so often, even though she stopped taking the drugs.

Fighting Illness in the Mind

Look in any bookshop and you'll see books on visualization and illness, on how you can beat even very serious illness by imagining your body fighting the disease. Not many of these claims are scientifically founded, but as you saw in the herpes example above, imagery can be effective in the mental battle against several different diseases.

If your immune system can be weakened by stress—which it can—then it should follow that alleviating stress might help to strengthen the immune system. There are many ways that we can learn to reduce stress, ranging from changing our jobs through learning to relax or signing up for psychological therapy. What is clear is that your state of mind has profound effects on how well your body can fight off the viruses, bacteria, and dangerous cells that moment to moment are gnawing at the cage of our immunity. Let's have a look at how mind and body interact in the fight against disease.

Immunity of the Mind

How precisely does our ability to fight disease depend on our state of mind? The immune system is an immensely complex system of defenses against infection. Central to this defense are some of our white blood cells, known as lymphocytes: B-lymphocytes, T-lymphocytes, and natural killer (NK) cells. All fight infection and disease, but NK cells are particularly interesting because they can kill certain virus-infected or cancerous cells in the body.

Our bodies have three main systems of government: the central nervous system, the endocrine system, and the immune system. All three are closely interconnected, so that changes in one usually produce changes in another. Our brains, for instance, rely on a vast array of chemical messengers to keep operating properly. These brain chemicals are needed to increase our chances of survival.

Suppose you were at this moment suddenly to hear a piercing shriek followed by a loud bang. You would immediately stop reading and hold yourself quite still as you listened to try to locate where the sounds came from and what they might mean. This vigilant state of your brain would be accompanied by a surge of brain chemicals—particularly one called noradrenaline—that would completely change your mental state and your mental functions. One effect of noradrenaline is to help make your sense organs more sensitive and tuned to the sights, sounds, and feelings that have captured your attention. This allows you to react quickly to danger and helps ensure your survival. Without this evolved survival mechanism, we would not have progressed this far as a species.

The immune system, however, also responds to many of these very same chemicals. Noradrenaline, for instance, can slow down lymphocyte activity and impede the destruction of virus-infected cells or cancerous cells by the immune system. This may be why severe or prolonged stress can lower immunity and increase vulnerability to disease: under such stress, our brains trigger the release of too much "fight-or-flight" substances such as noradrenaline over too long a period.

Stress, depression, and chronic loneliness can also affect illness more directly—through damage to the basic filaments of life, DNA. Damage to DNA and the body's strategies for repairing it are key to the development of cancers; states of mind such as severe stress can damage DNA at the level of the molecules themselves. This is just one of literally hundreds of ways in which states of mind can affect immunity.

But can we change these states of mind in the fight against illness? Yes, we can. You saw the example of the herpes virus above. There are many other examples of illnesses where psychological treatments have profoundly affected recovery. Take people recovering from heart attacks—myocardial infarction (MI). Several years ago, my research group in Edinburgh showed that stress management and relaxation training significantly improved recovery rates among people who had been hospitalized for an MI. Specifically, the patients given the psychological treatment were less likely to be readmitted to the hospital in the six months after their heart attack, and generally needed less medical care.

The mind-body connections are incredibly powerful. In the last chapter I showed how our brains make links between particular events in the outside world and certain reactions in our bodies and brains. These conditioned responses are learned, even to the extent that the body's physiological response to a drug such as heroin can be triggered by the sights, sounds, and smells that have become associated with taking the drug.

This seems to be possible for the immune response also. Rats in one study were injected with a powerful drug called cyclophosphamide, which suppresses the immune system. Mixed with the drug was ordinary saccharin. After repeated injections with this resistance-lowering cocktail, the rats were injected with the harmless saccharin alone. The result? Immunity plunged as if the cyclophosphamide were still being given. Links had been forged in the brain between the stimuli associated with the injections on the one hand and the body's plunging immunity on the other. Even when the drug source of this decline in immunity was withdrawn, the stimuli linked to it—particularly the injection of saccharin—triggered the body's immune collapse.

This happens in humans as well. Volunteers in Trier, Germany, were injected with adrenaline: in the short term, adrenaline produces an increase in NK cell activity. Each time they were injected, they also heard a particular noise and tasted a sweet sherbet in their mouths. This happened several times, until the two things were linked together. Then came the test. What happened when the sherbet and noise were presented on their own, without the injected drug? Sure enough, just as happened with the rats, the sherbet-noise stimulus now triggered an increase in key illness-fighting cells, the NK cells, in the volunteers' bloodstreams. The immune system had learned to respond to a completely innocuous taste and sound simply because this had been linked to being given a drug that actually did change immunity.

Another study showed the same kind of effect. You will probably at some time have had a Mantoux test to see whether you had natural immunity to tuberculosis or needed to be vaccinated against it. A little bit of tuberculin is injected under the skin of your forearm, and if you have had exposure to TB in the past, the site of the injection will become inflamed. A number of volunteers who had had TB exposure were given this test once a month over six months. The first five times, the tuberculin was taken from a green tube and injected into one arm. An inert liquid was taken from a red tube and injected into the other arm. Not surprisingly, the first five times, the arm injected with the tuberculin became inflamed, while nothing happened to the other arm. On the last test, however, the researchers put the tuberculin in the red tube. Even though it was precisely the same substance, the inflammation was greatly reduced. This was because the red tube had been linked unconsciously in the volunteers' brains to no response, and this association greatly damped down the skin response to the tuberculin.

These studies show how psychological factors—with imagery as a major component—have been severely neglected in Western medicine. Our immunity and our illness are hugely influenced by mental factors.

Are there particular times, places, or situations when you are more likely to develop a cold, or suffer some flare-up of an illness you are prone to? How about particular meetings or situations at work? Certain family obligations or encounters? Or maybe just particular states of mind or moods that come over you sometimes? Think about the last time you had a bad cold or similar minor illness. Can you pinpoint any things that were going on at the time that might have been linked to it?

We know that severe stress, depression, and isolation can downgrade your immunity and leave you vulnerable to infection. But your immunity can be affected—either improved or worsened—even when you are not in such obviously distressing states of mind. Changes in immunity can be conditioned to situations, feelings, events, and sensations inside and outside of yourself. If, for instance, you had repeatedly been given a treatment in a hospital that happened to lower your immunity, then it is entirely possible that the sights, sounds, and smells of a hospital could become conditioned stimuli that by themselves trigger a slight lowering of your immunity.

Some people might also learn to associate a particular house or living situation with sickness and lowered immunity, so that visiting that situation again could—theoretically at least—trigger the dip in immunity that allows that flare-up of a head cold. Maybe your stepparent's house isn't riddled with head-cold bugs after all; perhaps your brain and body have learned to respond to being there with a lowered immunity because you once went through a long period of being sick in that particular location.

Conditioned loss of immunity can be put to good use. Some diseases are caused by an overactive immune system that attacks the very body it is supposed to be protecting. Rheumatoid arthritis and lupus (systemic lupus erythematosus) are two examples of autoimmune diseases. Damping down the immune system is often a necessary part of the treatment in these cases.

In one study of lupus-stricken mice, the mice were injected with the immunity-suppressing drug cyclophosphamide, and each time this happened they were also given saccharin-sweetened water, rather like the rats previously referred to. As you might expect, the sweet water became a conditioned stimulus for the dip in immunity, so that it had much the same effect as the drug itself. An important consequence of this was that the conditioning allowed half the normal amount of the drug to be given to achieve the same therapeutic effect. Given that most drugs have side effects, this has big implications for clinical treatments for humans.

Even more incredibly, biochemical processes in the body known to be linked to damage to DNA can be conditioned in a similar way to that in which immune changes can be conditioned. In other words, the activity of molecules deep in the very filament of life can be influenced and shaped by mind and learning. No wonder that our state of mind has such a profound effect on our physical health.

I showed you in the last chapter how important the mind's eye was in conditioning. Conditioning is the type of learning that is at the heart of these remarkable mind-body interactions, and it seems very likely that the mind's eye has a part to play in influencing the body's capacity to fight illness. You saw how the mind's eye can re-create replicas of the events and stimuli that trigger bodily responses and so can help or hinder the formation of conditioned responses. Theoretically, then, these mental images may be able to trigger illness-fighting—or indeed illness-worsening—responses in blood and brain at the most fundamental levels of our physiology. Can we use the mind's eye to condition new bodily responses, or to reshape old ones, in such a way as to boost the effects of medical treatment?

There is so much quackery around, so many unfounded claims made by fringe medicine, that modern medicine is understandably very cautious about appearing to encourage its unscientific underbelly. Modern medicine is right to be skeptical. It has sometimes been said that the only effective drug is a new drug. By this is meant the powerful placebo effects that attend the hope and expectation surrounding new wonder drugs. These attendant psychological effects can have profound physiological effects on the sick body, most probably involving temporarily boosted immune systems.

Psychological treatments, including visual imagery, also benefit from such "new boy" placebo effects. Yet scientific scrutiny of such treatments has scarcely taken place, certainly in comparison with the scrutiny drug treatments get. So we must be extremely cautious in this question of what effects the mind's eye can have on illness. Nevertheless, there are some fascinating findings.

Conditioned Sickness

Chemotherapy treatments for cancer often produce unpleasant side effects such as nausea and vomiting. Roughly a quarter of the people going through this type of treatment develop what is called *anticipatory nausea and vomiting* (ANV). In other words, even before they are given the drugs, they show some of the side effects of them. What's more, these drugs unfortunately temporarily lower immunity through their effects on lymphocytes. Not surprisingly from what we know about conditioning, the stimuli associated with the chemotherapy—doctors, hospital, nausea—can themselves trigger a dip in immunity, even when the chemotherapy is stopped. These ANV symptoms add to the distress of an already distressing situation, and they don't respond to antinausea drugs in the way that the real side effects of the actual drug do.

Why do some people develop the symptoms of ANV and not others? Yes, that's right. People who are prone to becoming easily absorbed into what they are visualizing and imagining are the ones most at risk for developing this conditioned response. If you can easily "feel" yourself into states of heightened emotion, or readily and vividly re-create in your mind's eye the bite of the dentist's whining drill into the enamel of your tooth, then you would be more likely to develop these conditioned nausea responses if you were undergoing chemotherapy. Those of us with the ability to visualize vividly the bodily sensations and the surroundings in which treatments take place will be the ones in whom such imaginings can incubate and strengthen the development of conditioned responses.

The converse should also be true, however: Vivid mental images should be able to be used to undo some of these unwanted connections sewn into the brain. Indeed, ANV can be treated successfully by imagery-based therapies such as relaxation training and certain types of hypnosis.

Chili Peppers and the Mind's Eye

If you have ever had the misfortune to bite directly into a hot red chili pepper lurking in the depths of your Thai or Indian curry, you will be all too familiar with the fact that this plant contains a virulent agent—known as capsaicin—that causes inflammation of the soft membranes of our bodies. It is not just the mouth that capsaicin burns. If it is injected under your skin, it—not surprisingly—produces a burning sensation lasting for up to five minutes. It also produces a red flush on the skin called a *flare*. This flare is caused by the nervous system releasing chemical messengers called *neuropeptides*. This type of inflammation may be similar to some of what happens in autoimmune diseases such as rheumatoid arthritis—with the inflammation happening in the joints in that case.

A group of researchers in Iowa showed that the mind's eye can be put to good use in helping control even this apparently inevitable reaction to one of nature's hot and spicy substances. The people who volunteered for this study were given a pretest to see whether they could use mental imagery to change one sample aspect of bodily function—skin temperature. Try the following exercise to get an idea of the kind of test it was.

Sit down or lie down in a comfortable position. You can close your eyes if you want. Focus your attention on your left hand. Be aware of the feelings in it, and if you notice any tension, relax it. Now, using your brain's mental imagery machinery, imagine your hand becoming warm. Don't try to make it warm, just imagine that it is becoming warm. If it helps, you can imagine that your hand is resting beside a warm log fire. Or imagine that it is in a thick fur glove. Use whatever imagery appeals to you to imagine your hand warming up.

In the Iowa study, the researchers attached a temperature sensor to the middle finger of the hand while the subjects tried an exercise similar to this one. Many—but not all—were able to raise the actual temperature of this finger by using the mind's eye in this way. Only those people who could use imagery to raise the temperature of their finger by at least two degrees continued in the study. The fifty volunteers who had proved their ability to control body by mind to at least some degree were then trained in an imagery-based relaxation method. They were trained to further relax their bodies by using similar types of imagery to that used in the test. After they had demonstrated that they had mastered these skills, they were randomly allocated to three different groups for the chili test.

The first group practiced their relaxation for roughly half an hour before the chili substance was injected into their skin; the second were given stressful mental exercises; while the third—the control group—watched a yawn of a video entitled *The Bridges and Ferries of Iowa City*. After this period, the capsaicin was injected into the skin, and the degree of inflammation—flare—measured over the next hour.

The relaxation group showed a very different response from both the other groups. Twenty minutes after the injection, for instance, the control group's area of inflammation was approximately 66 square centimeters, while the flare on the skin of the relaxation group was only approximately 56 square centimeters. What's more, across all the groups, the size of the flare was accurately predicted by levels of stress measured before the injection.

This shows quite clearly how you can train the mind's eye not only to help control your emotions but also to help control the most basic biochemistry and physiology of your body. Imagery and relaxation can help you cope with that fiery curry, then. But what about greater challenges to the body? Can mental imagery help in more serious afflictions of the body?

Psoriasis in the Mind's Eye

The disfiguring skin condition psoriasis can be a source of immense discomfort, pain, and distress. The scaly lesions on the skin can be worsened by stress, though as a disease it is partly genetically caused. A group of researchers in Aarhus, Denmark, looked at the effects of mental-imagery training on the skin of psoriasis sufferers.

Fifty-one people who suffered from psoriasis were randomly allocated to either imagery, relaxation, or stress-management treatment on the one hand, or no treatment on the other. The imagery part of the treatment involved training the participants to imagine scenes involving sun and salt water on the skin. In reality, psoriasis can benefit from limited exposure to sun and salt water, and it seems that this is true in the mind's eye as well. The volunteers visualized sunbathing and bathing in the sea, picturing the scene in the mind's eye but, more importantly, imagining the sensations on the diseased skin as well. They also were trained to visualize the skin becoming pain- and itch-free—a painkiller of the mind's eye, in other words.

The Danish researchers used a laser technology technique to measure the blood flow in one of the most troublesome scaly plaques in each of the subjects. The imagery-trained group, who had visualized this plaque becoming pain-free, soothed by the caresses of sun and salt water, showed reductions in blood flow in the plaque that the control group did not. What's more, the people who were most successful were also best at visual and bodily imagery—they would have scored highly on the questionnaire in Chapter 4, for instance. They would also have scored highly on similar questionnaires related to how well one can create mental images of bodily movement and touch. In other words, those with a well-developed mind's eye could much better learn to control the disease in their skin than those who could not use mental imagery so well.

This was not just a nonspecific placebo effect of the psychological treatment in general. The researchers used the laser method to study the blood flow in the affected skin during and after normal relaxation training, and compared this to what happened after specific mental imagery focused on the particular area of skin. And true to what we know about the power of the mind's eye, the blood flow changes in the skin after imagery were very much greater after the specific imagery than they were after the more general relaxation.

In line with the changes in the skin measured by the laser method, the psoriasis improved in the treatment group compared with the control group on other measures. Of the imagery group, 91 percent showed an improvement in their psoriasis, as measured by a doctor who didn't know which group they had been in. In contrast, only 29 percent of the control group showed a doctor-assessed improvement.

The mind's eye can not only influence the very basics of the biochemistry of your skin, it can also be harnessed to help combat diseases that occur when that biochemistry goes out of kilter. It can be used to help psoriasis sufferers. Can it be used like this in any other diseases?

Migraine

Are you a migraine sufferer? Roughly 1 in 7 adults are, and even as many as 1 in 14 children are affected. While the exact way in which migraine occurs isn't known, it is associated with altered blood flow in the brain and in the brain's lining. These brain changes explain why migraine sufferers can suffer temporary partial blindness or visual disturbance, supersensitivity to sound, changes in the way the body feels, etc. A particular type of cell, known as a mast cell, is known to be closely involved in migraine. These cells, particularly in the brain's lining, release chemical messengers that can cause swelling of blood vessels, inflammation, and consequent pain. How active mast cells have been in an individual can be measured by the presence of a particular substance called tryptase in the urine.

Children between the ages of five and twelve with persistent migraine were divided randomly into two groups: a control group, and a training group who learned relaxation helped by imagery exercises. The children in the training group were taught to relax by creating visual images of pleasant scenes or objects of their own choosing. Examples included watching a koala or going skiing. The children were taught to use the mind's eye to visualize them, and thereby to reproduce the pleasant feelings that went with these things. The children were also taught to use body imagery to control headache pain—for instance, by visualizing the painful area becoming numb. To help them understand how mental pictures can affect what their bodies do, temperature and sweat sensors were attached to their fingers. By looking at a screen where their skin temperature and palm sweatiness were displayed, they could learn to increase these by using visual and other imagery.

This training not only reduced levels of the migraine enzyme tryptase measured in their urine, it also caused a big drop in the number of migraines these young children suffered. Visual imagery has also been successful in reducing other types of headache, such as tension headache. In other words, the mind's eye can be called into action to combat that most common of everyday pains. But what about more serious illnesses?

Coping with Surgery

If you have had surgery, you'll know well the pain and discomfort that typically follows. People who were about to undergo surgery of their colon and surrounding regions were studied at University College Hospital in London. Half of the fifty patients were given audiotapes before the surgery that contained imagery training about how to cope with the symptoms that usually hit people after such surgery. The tapes asked them to visualize feeling sick, for instance, and then to imagine thoughts and feelings of coping with this symptom, such as "you are occupying your mind by the thought that you are in control of the discomfort." Other symptoms such as pain, weakness, and dry mouth were also visualized, and again the patients asked to see themselves coping emotionally and in their thought processes with these symptoms.

Compared with the control group, who heard a tape that gave general information about the hospital, the imagery group needed significantly fewer painkillers after surgery, were in less pain and distress, and had lower levels of the stress hormone cortisol circulating in their blood. What's more, the imagery patients had spent only an average of forty-nine minutes listening to the tape. The effects of the imagery were pretty impressive.

The striking thing about the mind's eye is how virtually *anything* can be imagined—for good and ill. Not only can you visualize pleasant scenes to make you relaxed and less anxious—as the little children with migraine did—you can also visualize your own responses to imagined pain and distress.

Everyone has their Achilles' heel: some situation that you'd really prefer not to have to face, but face it you must. For some it might be a medical or surgical procedure. Others find the thought of meeting a particular colleague, boss, or ex-partner stressful. For others, just the thought of getting out of bed on a cold winter's morning is a downer.

Whatever your Achilles' heel, close your eyes and imagine yourself in that situation. Don't just picture it—*feel* and *hear* yourself into it. Try to summon the feelings and thoughts of dread that come with the situation. Now visualize yourself doing what the surgery patients did as they lay in bed awaiting surgery. Visualize yourself coping. Imagine your negative, tense feelings dissipating and turning into ones of calm neutrality and cool relaxation. In your mind's eye, see yourself behaving like some supercool, superrelaxed person who deals with this situation as if there is no history of angst or dread surrounding it. Imagine yourself immersed in a wry detachment as you enter the situation you fear.

This might not work the first time. But doing exercises in the mind's eye is no different from doing them in real life—you need *lots of practice*! The wonderful thing about the mind's eye is that this is a world over which you can have complete control. Unlike the real world, you can make things happen in your mind's eye, and visualize them in the most vivid and realistic detail.

But maybe you found it hard to keep control of the images—particularly if the situation you were visualizing was a very emotionally powerful one for you. That is understandable, and serious psychological problems require professional help. But for those of us with the more mainstream emotional wrangles of the normal spectrum of everyday life, our ability to learn to cope with them can be boosted greatly by practicing in the mind's eye.

As you saw in Chapter 4, imagery ability can be trained by practice. If you repeatedly try out imagery of coping in the way the surgery patients in London did, you have a good chance of improving your skills at coping. It's probably a good idea to start off by visualizing the less challenging situations first, and then work up to the most difficult ones. Give it a try.

Cancer and the Mind's Eye

There is no doubt that the mind has a part to play in the battle against cancer, but the only "magic bullets" for cancer will come from new treatments emerging from research in molecular medicine, not from psychology. Yet the practice of modern medicine is now fully briefed with the evidence—some of which I reviewed above—that the body's immune system can be boosted or depressed in its battle against cancer depending upon the mental attitude of the person.

If by any chance you have cancer, this book has absolutely no magic solution to offer you. There is no evidence that cancer can be cured by any psychological treatment on its own, and that includes mental imagery. On the other hand, there *is* evidence that your state of mind can *help* in the battle against cancer. To the extent that you find mental imagery helpful in improving your state of mind, then the mind's eye may have a part to play for you in the battle against the big C, or for your relatives or friends.

Some people have suggested that visualizing cancer cells in your body being destroyed by your immune system's natural killer cells can actually boost NK cell activity and combat cancer. The evidence for this is simply not available, though of course this doesn't mean that the exercise isn't worth trying—but only if you find it a positive and helpful experience and not a tiring or stressful business. One study in Denmark trained normal volunteers to visualize themselves "traveling through their own bodies," imagining different cells of the immune system. They had previously been taught about what these different cells were like, and were told to imagine these immune cells traveling through the body to attack an intruding virus. The study found that this visualization exercise did increase the level of NK cells in the blood of these volunteers, but the results were weak and need to be replicated.

While there are no studies showing that imagery on its own can directly affect cancer, there is quite good evidence that psychological treatment aimed at improving emotional state, sense of control, and fighting spirit can prolong survival to some extent in some types of cancer. In one study of group psychological treatment for people with malignant melanoma, even six years after the treatment there were significant differences between the treated and control groups. Of the thirty-four people in the control group, who had not received the psychological treatment six years earlier, ten had died in the interim. But of the thirty-four psychologically treated people, only three had died in the same period, a statistically significant difference. Of the control group, thirteen had had a nonfatal recurrence of their cancer, while this was true of only seven of the treated group. And what was the psychological treatment that produced such big differences in survival? Just one ninety-minute session of therapy once a week for six weeks, starting shortly after first diagnosis. The therapy aimed to educate patients about their health, reduce their fear, boost their coping skills, teach them how to manage stress, and improve their social support.

A comparable study was done with women with advanced breast cancer. Here, the women met regularly in a group with staff to discuss their fears of dying and to vent their emotions. They were also taught a simple relaxation method to help them cope with their pain. The researchers had not originally expected this therapy to prolong life: their aim was to reduce distress and improve coping. But in the light of claims about mind control and cancer, they looked at the data from their eighty-six women. While most of the women did die within three years, the psychological treatment bought them on average an extra eighteen months of life, as well as reduced pain and distress. (Fortunately breast cancer survival rates have improved a great deal since this study was done.)

It's clear, then, that your state of mind can influence, via its effects on your immune system, the course of even the most serious disease in your body. To the extent that imagery can help in improving your state of mind—and there is little doubt that it can—then the mind's eye could provide important weapons in the fight against disease. In the mind's eye, we can exercise a control over ourselves and the world that may be difficult to achieve in the real world. We can use imagery to get at least a degree of control over sometimes unruly brain and body systems. Once we do this, the external realities that limit our freedom and control can be altered by our *perception* of them. The mind's eye offers us a wonderful studio within which life's documentary can be edited. In a sense, the only reality that exists for us is our perception of it. In the mind's eye, that perception can be molded and morphed to an almost limitless degree. Few other mental faculties offer us this capacity for transforming the world in such an acutely realistic way.

Not surprisingly, imagery has proved very effective in helping people gain control of pain and illness, fear and anxiety. But this research has been badly neglected in Western medicine. One realm where the power of imagery has been much less neglected, however, is in cultivating the physical, athletic, and artistic prowess of our bodies. Which brings us to the next chapter—sports.

9

Visions of Olympus

Imagine you are a high diver at the Olympic Games. Visualize yourself walking toward the diving tower, feel your feet padding across the cold tiles. Imagine the goose pimples on your skin and the tightness of your swimsuit on your body. Your hands grip the cool iron rail of the ladder, your knee bends, and you feel the ridge of the lowest rung on the instep of your foot. You pull yourself up, and step by step, hand by hand, foot by foot, you steadily climb the ladder, aware of the changing perspectives and the gradually diminishing dimensions of the pool and its spectators. You are aware of the strain in the muscles of all four limbs at the effort of the vertical climb.

You reach up and this time there is no iron rung. This time there is the rough surface of the top board. Both hands are on it now and you lever yourself up onto it. You push yourself to a standing position and look up. You rock slightly on your heels as a faint wave of vertigo greets the taut, spread horizon. Feel the roughness of the board under your feet. Your hands are clenched, and you look down.

A mistake. The blue angles of the pool swim and blur with the pinprick faces of the crowd craning up to look at you from Lilliput.

You step forward, eyes up and ahead, and there is nothing except the dizzying sky surrounding you—no marker of your movement except the trembling feeling in your legs. You glance down and see the end of the board poised above the blue void. A single, black horizontal edge that you are walking toward. You keep walking, the remaining solid distance of the board shrinking to a gently bouncing square. Your body is rocking with the board and there is just one more step to the unforgiving edge.

You take the step and stop, looking straight ahead out into the absent infinity. But there is nothing solid where you can rest your eyes, and the swaying bounce worsens. You look down to fix on something solid, but there is only that edge and your toes curled over it—they too are undulating. Your heart is pounding as your mind claws the air for something solid to hang on to. But in looking down you have seen the handkerchief of blue water and its toy-town rim of human specks. Perspective gives your feet and the pool an appalling equivalence of size.

You swallow hard to try to stifle the dizzy, nauseating panic, and your arms swing to counteract the expanding to-and-froing of your body. These in turn excite the board to a more robust bobbing. Your toes are in spasm over the rim of the board and your body is now in an expanding loop of sickening swaying. In a second you will fall if you do not dive. You bend your knees, look down with one last gulp of fear, and jump off and down, trying to cast your body into an arrow behind your joined hands.

As I imagined myself on that board, I felt the dizziness and the pounding heart, almost as if I were up there. This was a scene that had almost all my mind-senses charged, right down to the taste of a dry mouth and the smell of chlorine. The sense of balance was particularly vivid. But maybe you have never been on a diving board—maybe this is a scene that you don't find particularly easy to visualize and feel in the mind's senses. If not, try some scene of your own. If you are a golfer, feel and see yourself on the fairway of your favorite golf course. If you are or have been a gardener, imagine yourself cutting a hedge with shears: feel the ache in your arms and the sweat on your forehead, smell the green and tangy freshness of sap bubbling from the severed stems. If you have swum in a warm sea, roll over and inspect the sky while the waves rock you.

In the mind's eye I am free. Free to live in the most vivid realism of any experience, any action, any scene in the world. This can be for good or for ill. I can face up to what scares me, or I can incubate corrosive fantasies. I can enjoy a soaring ski run while my body is crushed on the commuter train, or I can stoke up resentment by playing out in my mind imagined slights and put-downs.

There is no laboratory, no studio, quite like the mind's eye. Almost all of us neglect its potentially liberating—but sometimes oppressive—power. Not the world's top athletes, though. Few of them neglect it. For if they did, they would not be world-class athletes.

Excellence in the Mind's Eye

If you have ever seen Tiger Woods studying a crucial shot on the green, you might have wondered what he is doing when he crouches and stares fixedly at the ball. He is using his mind's eye to visualize the outcome of his putt. His father, Earl, taught him to use mental imagery while playing and training. Hunched over the ball, Woods is picturing the ball rolling into the hole. The American Olympic archer Janet Dykman uses a very similar mental training technique, visualizing the arrow going into the target seventy meters away.

Indeed the majority of the world's top athletes use mental

imagery, and most say that it is a vital technique for building up their performance. Among world-famous athletes who report preparing for competitions by "feeling" and "seeing" themselves acting out their sports routines are basketball player Michael Jordan, skier Jean-Claude Killy, golfer Jack Nicklaus, and figure skater Nancy Kerrigan. Olympic diver Michelle Davison takes long walks at night, rehearsing each dive in her mind's eye. She pictures the perfect dive, and feels very tired afterward from all the imagery. In her view, world-class athletes are all at much the same level physically: what distinguishes them is the mental realm, including how well they can use the mind's eye.

Luxury or Necessity?

But is this simply superstar fashion—a passing fad for imagery that doesn't really boost performance and build skills? Absolutely not. Not only does mental imagery help to build athletic excellence, having poor imagery may prevent otherwise physically accomplished athletes from reaching the top.

Researchers in the French city of Lyon have shown how imagery and performance are intimately linked. They studied international competitors in shooting and archery while they were concentrating just before making a shot, while they actually made shots, and thirdly while they imagined taking a shot in a competition. As I showed you in the last two chapters, imagining a scene or activity produces the bodily changes that would occur if you were really in that situation. Imagining yourself running, for instance, causes your heart to beat faster, just as it would if you really were running. The French researchers used this principle and measured heart rate, blood pressure, skin sweatiness, and other physiological markers while the archers and shooters went about their real or imagined business. Not surprisingly, the researchers found that when the sportsmen and women were concentrating prior to making a shot, they tended to be visualizing the shot they were about to make. Not only visualizing, of course, but feeling in their body and at their fingers and hearing in their mind's ear all the sensations that go with the shot. The same was true when they were shooting purely in their mind's eye.

The French researchers found that the pattern of heart and other bodily changes was uncannily similar in the three situations: in an imagined shot (whether during concentration preshot, or during armchair mental rehearsal), the heart rate would change at the same time as it did in the real shot. This allowed the researchers to calculate ratios of imagined to real physiological activity. In the perfect case, where imagery produced exactly the same pattern of bodily changes as the real thing, the ratio would be one. When these ratios were calculated and then compared with how the competitors actually performed, it was found that a close link resulted in a better performance. The more vivid and realistic the imagery, as measured at least by the bodily changes during imagery, the better the actual shots were.

Many other studies have shown a similar finding in many different sports. The best, most accomplished athletes seem to have a highly developed ability to replay very accurately their movements and skills in their mind's eye. Take volleyball, for instance. A key skill in this game is receiving a serve—that is, intercepting the serve of the other team and passing the ball to another member of your own team. You can measure very easily how well this is done by the accuracy of the pass to the teammate, and you can then study the effects of mind's eye practice on this accuracy. The French researchers who studied these players were able to measure how well they could play these shots in imagination by the pattern of heart rate and other physiological changes they showed during mental practice. The results were again clear-cut: those players who could replay in their mind's eye all the feelings, sights, and other sensations linked to the receiving serve were the ones who, afterward, showed real improvement in accuracy in their playing. Competitive swimmers also have a better ability to imagine key underwater movements compared with less accomplished swimmers.

If you play golf, then you will be able to improve your shots by practicing in your mind's eye. People have been shown to learn to putt more quickly and accurately if they imagine the exercise. Interestingly, however, visual imagery didn't work as well as two other types of imagery: kinesthetic (feeling of bodily movements) and auditory. In other words, if you imagine the movements in your body as you practice your putt, and use your mind's ear to re-create the strike of the club on the ball and the plop of the ball going into the hole, then your putt will improve more than if you rely entirely on visual imagery.

Young gymnasts with good imagery skills also are better at remembering and reproducing sequences of movements they have been shown. Being able to imitate through observation will clearly help in skill building. So it isn't surprising that the best athletes tend to be the best not only in the physical world, but also in the vivid simulations of the mind's eye.

To do this exercise, you'll need to have at least some experience of some type of sport or related activity. Maybe not for many years—it doesn't really matter. The point here is to imagine the different sights, sounds, and sensations linked to an athletic occasion. The more vividly a sportsman or woman can imagine these various scenes and feelings, the better he or she tends to fare in competitions.

Imagine your sports scene—whether it be around a pool table in a bar, at the aerobics class in your local health club, or at a high-level swimming gala. Golf, tennis, soccer, baseball . . . anything will do. Just pick some activity that you are familiar with, and visualize yourself in some scene linked to it.

Let's suppose that there are people watching—anything from a couple of friends to a crowd of ten thousand. Close your eyes. Can you feel your excitement with the eyes on you as you wait to perform? Do you feel a tinge of anxiety? Try to visualize yourself handling the anxiety and managing to stay calm. Now visualize yourself taking the shot, making the move—whatever it is. See how clearly you can re-create the feelings, sounds, and sights linked to it. Practice it a few times. If you find yourself making a mistake, slow down and repeat it, correcting the shot in your mind. Now imagine yourself performing it perfectly. Now make plans for the strategy for the game. Imagine yourself at different points in it. Now imagine you are tired, or that you lose concentration—you start to make mistakes. Picture yourself keeping your head, and keeping on going, refusing to be thrown off by the mistakes. Imagine yourself carrying on through the bad patch until your concentration and energy return.

See yourself in your mind's eye as confident and calm, highly focused on what you are doing, working through difficult moments without losing heart. Imagine that you look self-confident to the watchers, and visualize yourself as mentally tough, giving 100 percent to the task. Finally, picture the smiles and applause as you perform really well. Feel the eyes upon you and hear their congratulations.

It may all sound quite unrealistic, particularly if you have no special athletic skill. Personally, I am a pretty lousy athlete and have never won a sports event in my life. Nevertheless, when I run through these mental exercises, I really do get a sense of why it is that visualizing events can help to make them happen.

Mental training in the mind's eye, however, is possibly even more demanding than the real thing: it takes considerable concentration and effort to force your mind to re-create these events. With practice, though, as with all skills, we can get better. And if we improve our mental imagery, the research tells us, then we are a long way toward improving our real-life performance. But how does this happen? Why is it that sitting in an armchair imagining can actually build our real-life skills? What is going on in our brains?

Visualize yourself holding a heavy briefcase in your right hand. Keeping your arm straight, raise the briefcase out to the side and up to shoulder level. In your mind's eye, hold it there for a count of ten, and then slowly let your arm sink. Do this again. Lift, hold, and slowly lower. See how many times you can do this. Do you notice how your mental arm becomes tired? Did you slow down after several lifts because of the ache in the mental muscle?

Some people can mentally imagine such movements with great clarity, others less so. How tired you feel will depend on how vividly you inhabit your mind's eye. Just as the top sports performers tend to be masters of the mind's eye, so your ability to mimic real effort depends on your particular ability and self-training.

It takes some concentration to make your mind track the slow up and down of the mental briefcase. Sometimes the arm lifts and falls far more quickly than it would in real life. But if you practice it mentally, you will find that at least some of the time, your imagination creates an uncanny facsimile of the real thing.

Building Brain and Brawn in the Mind's Eye

The startling fact is that mental practice of this type can actually increase real-world strength. One study looked at the effects of mental versus real practice in tensing and relaxing one finger of the left hand. This mini-muscle building took place for five sessions a week over four weeks—a total of twenty training sessions. Half the participants actually did these exercises, while a second group just imagined doing the same exercises each day for the same total number of training sessions.

At the end of four weeks, the finger strength of each person was compared with that of a control group who did not train. The physical practice group's finger strength had increased by 30 percent, while the control group had only boosted their strength by a statistically insignificant 3.7 percent. In other words, if you do the equivalent of finger push-ups, you can build strength in the fingers—no big deal. But what happened to the people who just pumped iron in the mental gym? Their finger strength improved by 22 percent, almost as much as the physical training effect. In neither trained group did any other parts of the body show these increases in strength, which were quite specific to the fingers whether the training was physical or mental. Nor did the mental training work by building up the muscles in the fingers. In other words, the improvements in strength were caused by changes in the brain. And these brain changes were in turn caused by the repeated stimulation of the network of interconnected neurons controlling finger movements. This is good news indeed for the sluggards among us who would prefer to do our training on the couch rather than on the track or in the gym.

How does your brain achieve this? When you imagine yourself doing something, the brain switches on the very machinery that would be active if you really were seeing/doing what you are imagining. This has been shown using a PET brain scanner. Volunteers were studied as they imagined themselves moving a simple computer joystick. Researchers watched them to see which parts of the brain were switched on when the volunteers imagined moving the instrument. They compared this with what happened when the same people got ready to move the joystick, without actually moving it. What was found was that very similar parts of the brain "lit up" in these two different situations. In other words, mentally imagining a movement triggers much the same brain machinery as does preparing to make the same movement. It seems that imagining a movement is not very different from actually making the same movement as far as the brain is concerned. Only in the final stage—actually instructing the muscles what to do when—does the brain call into action extra centers that are not involved during imagination.

This is why the world's top athletes almost all use mental imagery to train and tune their skills and strength. It is as central a part of training for them as actual physical training. If you are a horseback rider, golfer, tennis player, pool shark, swimmer, runner, jumper, darts player, bowler, or anything else, then you can, if you want, improve your skill while lying in bed on a Sunday morning. This is not speculation. This is hard, scientific fact. Lying with your head comfortably on the soft pillow, your brain has to use almost all the neurons and circuits that it would use if you really were out there practicing your sport. And if you manage to discipline your mind's eye to go through your training routines, you will strengthen and refine these brain circuits just as you would if your body were actually joining with your brain in this enterprise. To repeat, physical strength is in both body and brain: physical skill is largely in the brain. It follows that to hone your skills, much of the time you don't really need your body!

Take this study that plotted brain changes while volunteers learned a particular physical skill: a one-handed five-finger exercise on the piano. For five days, two hours a day, they practiced this exercise and, just as has been shown in much other research, the area of the brain activated by this sequence of movements gradually expanded as more brain cells were recruited through practice into controlling this increasingly fluent and skilled movement. Just to make sure that the brain changes were really caused by the specific learning of the note sequence, another group of people played one-handed piano for the same period, without practicing a particular exercise or sequence of notes. Their brain areas expanded far less than those of the specific exercise group, while a control group who didn't practice at all showed no change in the brain areas controlling movement in that hand.

Most interesting of all, however, was how a fourth group of people fared. These were people who trained in the mind's eye, mentally practicing the five-finger piano exercise for the same length of time each day for five days. What happened to their brains? They showed similar changes as the brains of people who really did practice the exercise! Which goes to show that visualizing yourself training in the mind's eye improves strength and performance by physically changing the brain.

Top athletes spend years building up wonderful networks of brain connections; it is in these assemblies that their talent is stored. Maintaining this pattern, however, needs constant rehearsal; otherwise the trained pattern of brain connections will wither from lack of stimulation. But this rehearsal can happen either in the real world—or in the virtual world of the mind's eye.

So the next time you are laid up with a cold, you can console yourself with a mental training session. But it is important while doing this that you use all the virtual senses—the sense of movement and other bodily sensations especially. We know this because of a study that compared brain activity in people who were *watching* hand movements with their brain activity when they *imagined themselves* moving their own hands. Watching someone else moving a hand triggered just the visual parts of the brain at the back of the head. Imagining making the movement themselves, on the other hand, lit up the movement areas of their brains. So you wouldn't get fitter if you lay in bed picturing yourself sprinting down the street. But you would probably get a little fitter if you *felt* yourself doing this. Fitness resides partly in the brain connections, and you don't exercise the brain connections of the movement areas when you *watch* yourself doing the exercise in your mind's eye. Mental training requires the mental body to be active. To push home this message of how the mind's eye can faithfully mimic the real world, do this exercise.

Imagine that you are about to write your name, address, and telephone number on a piece of paper. Before you start, set a timer so that you can find out how long this mental exercise takes. Take care to imagine your hand movements precisely. Now start the timer and begin. Make a note of how long it took you. Now find a piece of paper and do the same exercise in reality. Again time yourself, and compare the two times. If you were to do this several times, you would find a close relationship between the time it takes to complete the mental task and the actual one.

You will have done the last exercise with your preferred hand—in most cases the right hand. Now do the exercise again, but this time with your nonpreferred hand—in most cases your left. Time yourself in the mental and the real conditions again. You will not be surprised to find that it took longer in reality to write your name and address with the left hand (or right if you are left-handed). But if you did as the people taking part in another study did, then you may be surprised to find that mentally writing with your left hand also took longer than with the other hand. In other words, mental simulation followed very similar rules to those applying in the real world.

The same happened when volunteers imagined writing text in large versus small letters. Just as writing in large letters took longer in reality, so too did the mental act of writing in large letters. And the same was true for walking—imagining yourself walking ten yards took longer than imagining yourself walking five yards. Furthermore, the time taken to walk the imagined ten yards was very close to the time taken to actually walk ten yards.

It's not just in sports that you can use the mind's eye to improve your learning. Surgeons can improve their operating skills and physicians can improve their techniques of internal examination by practicing them mentally. Musicians can practice even while their instruments are in the baggage hold of the plane as they fly to a concert. The world-renowned concert pianist Glenn Gould practiced very little physically in the last part of his career. Instead, he would read a score and mentally practice it several times before going on to make a recording of the score he had practiced only in his mind. If you are a musician, try this. Try practicing a piece or a scale in your mind's ear and mind's fingers. You will find that you improve with this mental practice.

Rehabilitation in the Mind's Eye

If mental practice helps improve skills and boost learning, then should it not be possible for people who have suffered a stroke or other types of brain damage to use the mind's eye to help in their recovery? Yes, it is.

One research group in Dusseldorf, Germany, used mental imagery to train nine people who had partial paralysis of one side of their body following a stroke or (in one case) an injury following a blow to the head. They were trained to reach for a glass with their semiparalyzed hand, initially with the help of a physiotherapist, who helped guide the arm and hand in the correct sequence of movements. Then the patients had mentally to practice this movement in their mind's eye over and over and over again, taking care to try to re-create the sensations of movement and the feelings in their arm as they did so. This training led to significant improvements in the quality of the movements made by these patients. In other words, for some people at least, their therapy time can be greatly increased if they use their mind's eye for rehabilitation.

But imagining movements like this depends on at least some of the brain circuits that control movements still functioning to some extent. Without any of the movement apparatus in the brain, it would be difficult to re-create the sensations of movement—except perhaps through visualizing what the movements look like, or imagining some of the internal sensations linked to them. Movements, however, are controlled by several different circuits in the brain, and it is likely that in many people there will be at least some chance of imagining the surviving brain circuits into life and giving the damaged brain the possibility of at least some improvement in functioning through mental practice.

The mental gym is always open, it seems. The better you can visualize, feel, hear, and sense in the mind's imagination machine, the better chance you have of developing athletic, musical, and other physical skills. What's more, you can practice and train these skills without ever leaving your bed, and the result will be better performance and strength. But are there any other ways into the mind's eye other than by effort and practice? Are there any shortcuts to vivid imagery? Let's take a look at one about which many incredible claims are made—hypnosis.

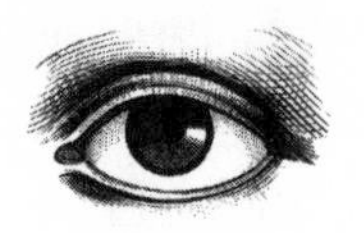

10

Hypnosis: The Imagery Game

The professor wanted to find the most hypnotizable students in the class. "Hold out your hands in front of you, about a foot apart, palms facing inward," she told them. Their eyes closed, the class obeyed—a strange sight for anyone passing in the corridor outside. "Now imagine some force pulling your hands together—it could be a magnet in each hand, or an imaginary rubber band pulling them together. As you imagine this, your hands will slowly move together, pulled by the force." The professor looked at her watch and then at her class—ten seconds had passed, and she saw that about half the students had moved their hands closer. Some students' hands were touching—drawn together by the imaginary magnetic field. Others' hands had not moved at all.

Next she suggested to the students that they could hear singing. "Who actually heard the singing?" she asked. "I mean, who actually heard it as if it were outside your head in the world, rather than just imagined inside your head?" From the class of four hundred students, eight slowly raised their hands. They were the elite—that minority of the already highly hypnotizable who can actually hallucinate when it is suggested to them under hypnosis.

But were they really hallucinating—actually hearing the music outside their heads in the world? Or was it just another form of normal imagining, where the sounds were inside their heads—the way we can all imagine music or voices? This is the age-old problem: hypnosis as quite different state of mind versus hypnosis as good amateur dramatics linked to a vivid imagination. Well, with these eight students' agreement to go into a brain scanner, we were about to find out.

The students were enrolled in an experiment where they listened or imagined listening to a simple taped sentence—"The man did not speak often, but when he did, it was worth hearing what he had to say"—while their brain activity was measured. First they listened to the sentence. Then they were asked to imagine hearing this sentence as vividly as possible. Finally, they heard only the sound of the tape recorder being switched on and they had to hallucinate the sentence being read, just as they had hallucinated the singing in the classroom.

Six of their classmates were also enrolled and followed the same instructions in the brain scanner. Although just as hypnotizable in all other respects as their hallucinating classmates—their hands were just as likely to have clamped together under suggestion, in other words—they had not been able to hallucinate sounds under hypnosis. The question was, did their brains and those of their hallucinating classmates behave any differently?

Yes, they did. When the hallucinators were hearing the voice outside their head, a distinct part of the right half of their brains was particularly active. This brain area is called the anterior cingulate, *and it was not active in the nonhallucinators. Nor did it "light up" when the hallucinators merely imagined the sentence as vividly as they could; only under hypnotic suggestion to hallucinate did this brain center become active. The students also had to give a 1 to 10 rating to how much outside their heads the voice seemed, and another 1 to 10 rating to how clear the voice was. The higher these ratings were, the greater was the activity in the right anterior cingulate part of their brains.*

So the students who could hallucinate sounds under hypnosis weren't just good playactors. Their brains could respond to hypnotic suggestion in a very particular way that located sounds outside in the real world, rather than inside their own heads.

How Absorbed Do You Become?

Remember in Chapter 7 I asked you how *absorbed* you tended to become in various sights, sounds, thoughts, or other sensations? A piece of music, a play, a movie? Maybe just a particular effect of sunlight on a choppy sea, or the swaying of a tree in the wind? Or the face of someone you love, as they speak, or laugh, or sleep? Absorption means your attention is completely and wholly focused on one single image, event, or thought. So fully occupied is your attention that there is none left to pay attention to yourself: We lose our self-consciousness when we are absorbed. We also lose our guile, shedding self-protective layers of self-awareness. Young children are absorbed much of the time, and to watch them watching a spectacle is to see the brain's attention systems focused on a single point as hot and bright as gathered sunrays through a magnifying glass. And like the focused sun, attention can burn its way into the heart of a scene, a sound, a smell, a feeling.

Choose a sight, sound, smell, or other sensation that you really love. A piece of music, perhaps, a tree or flower, the smell of cut grass? It could be touch—the feel of someone's fingers massaging your shoulders. It doesn't matter what it is. Now relax and settle yourself down for five minutes to absorb yourself in this. Set a timer if necessary, so you don't worry about the time. Relax into the sensation, the sight, the sound, or the smell and let your attention burn into it like the sun through a magnifying glass onto paper. Imagine your consciousness focused on this one thing. Don't worry if other thoughts or sensations come into your mind. If you become aware of them, just bring your attention back to the focus of consciousness for this five minutes. In short, try to lose yourself in this focus.

The phrase "lose yourself" tells us most of what we need to know about the phenomenon of absorption. Did you manage to "lose yourself" for even a few seconds during this exercise? If not, try it again sometime. But don't try *too* hard, because anxious effort sabotages attention and makes it difficult to become engrossed, absorbed.

Adult audiences can be absorbed by a great orator. A glance at old newsreels of Hitler at rallies will show the uplifted, spellbound faces of the audience. Adolf Hitler was said to have a certain hypnotic quality about him, and certainly the propaganda shots of his filmmaker Leni Riefenstahl give the impression of an audience appearing to be in some sense of the word hypnotized. Hypnosis, it seems, may be a particular form of intense absorption—a sort of self-forgetting immersion of all one's attention into one small focus on the world or in your thoughts. The more prone you are to becoming absorbed in thought or sensation or fantasy, the more hypnotizable you are likely to be.

Have you the capacity for absorption? Answer these questions: Can a sound—say of a voice—become so engrossing for you that you just keep on listening to it? When watching a play or movie, do you sometimes become so involved in the emotions of a character that it feels as if you have "become" that person for a few moments? Can you recall past events in your life with such striking vividness that it is almost as if you are reliving them? Do you sometimes daydream or fantasize so vividly that the images hold your attention the way a good story or play does? Do you sometimes forget about yourself and become engrossed in a fantasy that you are someone else? Could you, if you wished, picture your body as being so heavy that it was immovable?

If you answered yes to several of these questions, then you have a strong capacity to become absorbed. Apart from being more hypnotizable than average, you will also tend to have a well-developed ability for mental imagery. You may also be somewhat more creative than average. You have, in short, the ability to slip the reins of the "cool web of language" and enter into the realm of sensation and imagery. Hypnosis seems to be a technique that helps people to do this. Maybe we can understand more about the mind's eye if we study what happens to the brain under hypnosis.

What Is Hypnosis?

Hypnosis seems to depend quite heavily on some of the brain circuits that you need for mental imagery. When you watch a stage hypnotist telling a woman on stage to reach out for the shiny red apple in front of her, she may well be "seeing" that apple, because her brain has activated the same circuits in her visual cortex that would be active were she actually seeing an apple. Stephen Kosslyn—the doyen of imagery research—showed this in a study where volunteers were hypnotized and told they could see color in what was actually an all-gray pattern. Their brains were scanned as they looked at these gray patterns, and sure enough, the color-processing regions of the visual brain became active in line with the suggestion that they could see color in the gray. When the same volunteers were told under hypnosis that they could see only gray patterns in a color picture, these same color regions became less active again. In other words, hypnosis really can change the brain in a very powerful and particular way. The person hypnotized to "see" the nonexistent object or color really can see it.

So while some of the stage hypnotist's volunteers may simply be playacting on the stage, many will actually be experiencing the sensations, thoughts, and feelings suggested to them. That is why stage hypnotism should not be encouraged. Hypnosis can produce temporary brain changes as powerful as any drug, and drugs can't be given on stage for fun.

One of the classic hypnotist's tricks is to suggest to subjects that they won't be able to move an arm, leg, or some other part of the body. A recent study of such hypnotically induced paralysis showed that the same part of the brain that was involved in the hallucinations under hypnosis was also involved in hypnotic paralysis. Peter Halligan of the University of Cardiff in Wales and his colleagues hypnotized a volunteer, suggesting to him that he would not be able to move his left leg. Sure enough, he wasn't able to move it, and brain scanning showed that the anterior cingulate part of the right hemisphere of his brain was more active than when he simply prepared to move his left leg, unhypnotized.

This same brain area—the right anterior cingulate—has been shown to be a key area for hypnosis in yet another hypnosis study, this one on pain control, carried out by a team of researchers in Montreal. When volunteers were simply hypnotized into a relaxed state, they did indeed become relaxed, somewhat detached from reality, and with an altered sense of time, space, and self. Afterward, they told the researchers that they were conscious of much more visual imagery than normal. In line with these feelings, the biggest change in brain activity during hypnosis was in the occipital lobes—the vision centers of the brain—which you saw earlier in the book are so important for visual imagery. But activity in these image-generating brain centers went hand in hand with increased activity of that key hypnosis-related center in the right hemisphere of the brain—the right anterior cingulate.

This hypnotic trance also turned *down* the activity in some other brain regions. In particular, farther back in the right hemisphere, in the lower part of the parietal lobe, brain cells fired much less strongly during hypnosis than normal. It is in this part of the brain that we keep maps of the space around us, of our body structure, and of the link between the two. You saw in Chapter 3 what can happen when these centers in the right hemisphere are damaged—people can lose the ability to create images of the left side of space and the left side of their bodies, as well as having difficulties in forming mental sketches of the structure of space around them.

When putting people into a trance, the hypnotist tries to do something similar—albeit temporarily. Under hypnosis you want to help people detach themselves from their body and from external space to some extent, and—sure enough—the brain machinery whose job it is to update regularly the maps for where you and your body are in space is ratcheted down under hypnosis: hence the dreamy detachment and altered sense of self, space, and time that is the hallmark of the hypnotic trance.

Pain is the enemy of this pleasant and altered state of the brain. When the Montreal group had their volunteers put their hands into uncomfortably hot water, the dream-image–generating visual centers of the brain switched right down, and the space and self-locating centers in the right parietal lobe jumped into action again. Anyone who has experienced pain knows how commanding its presence is on your attention, and how its precise location in your body cannot be ignored. Under hypnosis, however, you can reduce pain through various types of suggestion. The Montreal researchers gave the hypnotic suggestion to the volunteers that they would not find the sensation of their hand in hot water unpleasant, and—sure enough—these suggestions worked and the hot water was much less uncomfortable. But while the general hypnotic relaxed state worked by switching on the right hemisphere and visual centers described earlier, the suggestion part of the hypnosis activated more centers in the left half of the brain. This was related probably not only to the decoding of instructions by the language centers of the left hemisphere, but also to the brain, particularly in the left frontal lobes, changing the *interpretation* of these gnawing painful sensations in the hand from more to less unpleasant.

So while the hypnotic trance depends on the right hemisphere and visual imagery changes, hypnotic suggestions demand the help of the "cool web of language" in changing our perception of sensations. So powerful can hypnosis be in changing how we respond to sensations that amputations have famously been performed under hypnosis. In 1842, for instance, a surgeon by the name of Ward reported amputating the leg of a patient while the patient was hypnotized. The evidence was consistent in suggesting that the operation was performed painlessly, but the Royal Medical and Chirurgical Society refused to accept this. Eight years later, on the basis of a rumor, the eminent physiologist Marshall Hall had the paper struck from the society's records, saying that the patient had subsequently admitted to falsely denying his pain. This was in spite of the existence of a signed declaration by the patient stating that the operation had been painless.

There are many more recent anecdotal accounts of major surgery being performed under hypnosis, and hypnosis is used by some dentists throughout the world. In recent years, hypnosis has usually been combined with lighter-than-normal anesthesia where the surgery is major, and there are many, many studies showing that hypnosis can help greatly in the control of pain.

Ernest and Josephine Hilgard, pioneers in the use of hypnosis for pain control, describe in their book *Hypnosis and the Relief of Pain* a forty-two-year-old woman who had unrelenting pain in her right arm due to bone cancer. Josephine Hilgard induced hypnosis via brief relaxation and getting the woman to count her way slowly into the hypnotic trance. Then she was asked to travel back in time to a particularly pleasant occasion she remembered, being by the sea on a beach, and swimming in the water. As the hypnotic state deepened, the hypnotist induced her to feel her left—nonpainful—hand becoming numb and anesthetized, so-called glove anesthesia. Gradually she was able to imagine this hand becoming cold—for instance, by using her mind's eye to imagine plunging her hand into a bucket of ice, or building a snowman with it.

Hilgard helped this imagery along by telling her hypnotized patient that a very cold hand eventually becomes so numb that the sensation in it diminishes until it feels no pain. Then, having helped her change the sensations in her left hand, the hypnotist turned to the right arm and hand, where the pain was constant, and barely tolerable. She suggested to the woman that she imagine herself gradually pulling a long evening glove onto her right hand with her numbed left hand. Using this visual and tactile mental picture, she encouraged the patient to visualize slowly and carefully, finger by finger, this long numbing glove being slowly rolled onto the pain-racked arm, up over the elbow. The woman gradually felt the pain diminish in that arm, to the extent that she could now use it much more freely than before. What's more, she learned to hypnotize herself so that she could use this technique at will without the help of the therapist.

Anecdotal case reports like this are not scientific evidence.

But the actual scientific evidence now—for instance, some of the brain-imaging studies above—shows clearly that hypnosis can be a very, very powerful method for controlling pain. And it seems to control pain by changing how the brain responds to it, just as pain-controlling drugs do.

Hypnosis seems to work in part by boosting the brain's imagery capacity. We know that imagery involves profound changes in brain activity, so it is no surprise to find that hypnosis has similar—probably even more dramatic—effects. And given that pain exists only in the brain, then it is clear that hypnosis and imagery can—and do—help reduce pain.

The glove suggestion above is only one of an infinity of possible suggestions that might have been made to the woman with the bone cancer. Hypnotic imagery is virtually unlimited in the scope of the states and experiences that it can evoke. Josephine Hilgard describes, for instance, how her patient gradually came to be able to visualize herself floating, to the extent that—only in her mind, of course—she found herself floating out of bed. She found that this out-of-body experience helped detach her from the clutches and discomforts of her illness.

Hypnosis and guided imagery work with children too. In one study of fifty-two children who had surgery, the twenty-six who were randomly assigned to hypnosis and guided imagery not only had much less postoperative pain, they were also able to leave the hospital significantly earlier than the twenty-six children who did not receive this training of the mind's eye. The same has been shown in adults, and during painful cancer treatments, hypnosis and imagery training significantly reduced pain in adults receiving bone-marrow transplants in a major cancer hospital in Seattle. There have been many similar studies with both adults and children enduring the pain and discomfort of chemotherapy, radiotherapy, surgery, and other treatments for cancer and other painful diseases. Hypnosis and imagery are clearly effective in helping control pain, anxiety, nausea, and vomiting.

Whether in health or illness, imagery gives us all a glorious freedom to travel in time and space, no matter what the actual state of our physical body and environment is. One in three of us will suffer cancer in our lifetime, and I know that if I do, I will certainly make sure I assiduously practice imagery as a means of controlling pain and discomfort. And I will probably seek out a properly qualified hypnotist—carefully avoiding the many charlatans who haunt the vulnerable—to help me train up the vast potential for imagery that most of us have.

Anything that cures can also harm, and this is no less true of hypnosis than of drugs or surgery. Many unfounded claims are made for hypnosis by therapists eager to make money. Most of the effects of hypnosis are temporary, and if they are to last, it is because the person learns to control his or her own brain in order to change it.

As you saw in Chapter 6, visual imagery of imagined situations that never happened can be so compelling that children and adults can come to believe that they actually did happen. Because hypnosis is such a powerful technique and because it relies so heavily on visual imagery, it is not surprising to find that "recovered memories" of so-called alien abductions, firmly believed by the "victims" to have occurred, are often first "recovered" under hypnosis. It seems that the fantasized abduction is so vividly visualized that in the person's memory it becomes mixed up with real events.

Hypnotists are not possessed of magical powers. Anyone with the self-confidence and the training could hypnotize another person. What is happening under hypnosis is that one learns to release certain capacities in one's own brain that temporarily change its state. Imagery plays a large part in this—and the hypnosis research shows us just how much unused potential all of us have to escape the "cool web" and use the mind's eye to its full capacity. Most of us can learn to use the power of our mind's eye without ever visiting a hypnotist. Because it is not the hypnotist who changes your brain—it is *you* who does. Sometimes, however, vivid images come to us unbidden—particularly in that unfathomable world of dreams. Let's look at these in the next chapter.

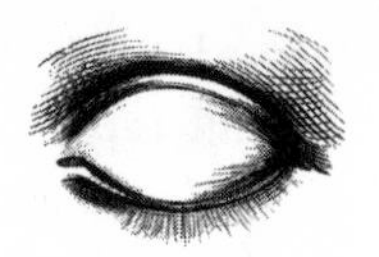

11

Dreams and Images

It is in dreams that our mind's eye fully slips the leash of language and roams through the galleries of impossibility. The things we fear, those we long for—the two sometimes absurdly mixed—flash before our eyes in a pictorial pageant that eludes the waking mind. This alternative, topsy-turvy world in which we spend a sixth of our life is a textbook illustration of the mind released from the "cool web." In dreams we see fear and longing running loose, leaving the jockey, language, lying winded and concussed behind the high fence of sleep.

Dreams and Sleep

The more vividly you can see, feel, and hear on your mental screen by day, the more likely it is that you will tend to remember the vivid dreams that haunt your sleep. There are several different stages of sleep, and it is the one that happens later in the night—rapid eye movement (REM) sleep—that is linked to the kind of vivid dreams we tend to remember. During REM sleep, you make numerous flickering eye movements behind your closed eyelids, and your body is still. If you wake someone during REM sleep, they will be twice as likely to report that they were having a dream than if you wake them in other stages of sleep, which tend to happen earlier in the night. You also dream during non-REM sleep, but the dreams are different. Non-REM dreams are like our normal thought and dwell more on the day's mundane worries—the brain's sober prefects pacing the corridors of the mind, brows furrowed with institutional concerns.

REM dreams, on the other hand, are more vivid, pictorial, and multisensory. They also involve the sensation of movement and action and are more emotionally charged. They are much less linked to real, waking life—in their bizarreness, for instance. Giddy and unstable drama queens and kings come out to flaunt their gaudy glad rags in surreal and flagrant disregard of rule book and rationality. The thing about these overdressed dramatists and their zany stage sets is that they really believe that this is all true—that this is reality. Think yourself of the relief of waking up to find that the giant mongoose is not about to eat your best friend. But before you woke up, you were desperately afraid for your friend—this was very, very real.

Why are we so completely duped in REM dreams? One reason is that our brain chemistry changes during REM sleep, with the brain's chemical messenger acetylcholine boosted up beyond normal levels, while its sister chemicals noradrenaline and serotonin have their taps turned off. These and other sleep-induced changes in the brain are linked to a switching off of the brain's chief prefect—the frontal lobes—as well as most of the other prefects in the parietal lobes.

These brain regions being the seat of self-awareness, it is not surprising that the school drama society believes the play to be real as they strut their stuff without supervision or constraint. This is especially true given that the emotional heat of the stage is turned up via big increases in the activity of the brain's emotional systems in the amygdala and limbic areas. Interestingly, the same brain area deep in the middle of the frontal lobes that "lights up" under hypnosis—the anterior cingulate—also increases its activity in REM sleep. It should also be no surprise to find that similar brain areas "light up" in REM sleep as do with wakeful imagery. When you see that mongoose poised to eat your friend, you can be sure that many of the same regions, toward the back of your brain, are in action as would be the case if a mongoose really were there.

Similarly, if you feel yourself flying, falling, or running (I pray that none of my more Freudian-inclined colleagues read this book), then the brain systems responsible for balance, flight-like arm movements, and pedaling legs will be firing in much the same way as if you really were engaged in these activities. What's more, if in your dream you are running—say—from your car to the front door of your house, there will be a close correspondence between the number of seconds the neurons in the motor/running parts of your brain fire and the subjective duration of the dream run.

Like visual and other types of imagery, then, the images of what we see, hear, feel, or do seem to be produced by cranking up the machinery that would be used to engage in these activities in real life. The main difference is that these images run free of the self-conscious monitor of the frontal lobes, hence their unruly behavior and bizarre content.

Fortunately for us, we are effectively paralyzed during REM sleep, meaning that we don't act out the strange things we do in the world of dreams. In rare cases, however, this useful paralysis fails, leaving open the channels between dream image and final motor control, opening the pathways from brain to muscle. One man famously, while fast asleep, held his wife in a headlock and, making running movements with his legs, shouted, "I'm gonna make that touchdown!" He then tried to throw his hapless wife's head toward the foot of the bed. When he woke up, he remembered a dream in which he was running with a football for a touchdown.

A patient of mine was perplexed to wake up and find his bedroom devastated. The wardrobe had been toppled, chairs were overturned, drawers spilled out on the floor, and lamps smashed. Assuming that he had been burgled, he called the police, who came, took a statement, and left, puzzled that they could not identify where the burglar had made his entry. As my patient relaxed that evening, recovering from his trauma, sipping a restorative glass of wine, he suddenly remembered the dream he had had. He told me, "I suddenly remembered a vivid dream that I was traversing this very difficult cliff face, with lots of obstacles and big rocks that kept slipping from under me . . ." So he had—minus ropes and belays, one assumes—traversed the north face of his bedroom as wardrobe, drawers, and lamps crashed into the dream precipice below him. Sheepishly, he phoned the police station to tell them to call off their inquiry. Like the man touching down his wife's head, his brain had gone into the REM dream state in all but one crucial detail—the "off" switch for the muscles that stop us acting out what we dream was not activated.

But I Can't Remember My Dreams . . . or I Wish I Could Forget Them

There may be another reason why our REM dreams are so bizarre, fragmented, and unpredictable. This is that the neurochemical changes in the brain during REM sleep make it effectively amnesiac—within the dream itself as well as after you wake up. In other words, even while you are dreaming, you are forgetting what you have just dreamed, leading to ridiculous shifts of scene, plot, and character in the surreal mélange that is the dream world.

It may be possible to remember more of your dreams if you acquire the discipline of writing down what you dreamed as soon as you wake up. People who are plagued by recurrent nightmares can in some cases get rid of them by writing them down and by thinking about them in a more relaxed state than applies in the torrid emotional climate of the dream. If you can also rehearse in your mind's eye the bad dream with a different ending, this can help beat the nightmares too.

You don't have to have had a traumatic experience to have recurrent nightmares, but a trauma usually causes recurrent nightmares. In one study of no fewer than twenty-three unfortunate children who had been kidnapped and buried in a truck trailer, all had nightmares about their experiences. Eight out of ten children who had witnessed their mothers being raped went on to have nightmares, while six out of ten children who had been in a school playground that came under sniper fire had recurrent bad dreams about that traumatic event. And nightmares are almost universal among Vietnam veterans.

Children who have never had traumatic experiences tend to have dreams filled with wild animals and—often—fear of attack by them. While this may have to do with the books, videos, and TV programs that we typically expose our children to, some scientists have argued that this is evidence for biologically inherited fears of wild animals that are no longer a threat in our lives but were when our brains were evolving. Certainly our fear of snakes is likely to be inherited from early evolutionary times, which explains why so many city dwellers with zero chance of ever meeting a snake are still irrationally frightened of them.

Dreaming, it is also argued, is a biologically useful process. It allows us to rehearse the dangers we are likely to face during the day, as well as to rehearse our responses to the threats—whether fight or flight. After we have had a traumatic experience, the argument goes, the recurrent nightmares are our brain's attempt to rehearse an escape from the trauma to help us survive such an event in the future.

In a study of Palestinian and Jewish children's dreams, aggression appeared in almost all the dreams where there were interactions between people, and only a tiny proportion of interactions between people involved friendliness. Jewish children dreamed of Palestinian attacks and explosions, while Palestinian children living in refugee camps dreamed about brutal physical aggression, arrest, beatings, and killings. Each group of children dreamed of being attacked by adults from the other side to the extent that the conflict between them was even more awful and destructive in their dreams than it is in reality.

Daydreams are often different. In daydreams, we tend to have images about what might happen in the future, in contrast to the past-haunted dreams of night. An exception may be in traumatized adults and children, who even while awake might be haunted by the brain's replaying of the terrifying experiences they have undergone.

Learning in Dreams

REM sleep may involve one particular type of stimulation of brain circuits that helps consolidate and strengthen the new learning and memory that the day's experience has half-crocheted into your brain. Evidence for this comes from research indicating that during REM sleep the weave of the crochet is tidied and tightened, leading to overnight—effortless—learning. This research was based on a visual perceptual test. This test required people to detect small differences in a complex design. If tested after a night's sleep, the volunteers were found to be better at the task than they were at the end of a session's exposure to the test the previous day. In other words, their brains were learning as they slept, tightening and tidying the previous day's crochet work of learning and improving performance the next day.

The researchers then looked at what happened to this overnight learning if they interrupted REM sleep by rather unkindly waking up volunteers (between twenty and sixty times during the night) whenever they started to show signs of REM sleep. Sure enough, when deprived of REM sleep, these people did not show the overnight learning effect. And this was not just because of tiredness—for when they were wakened up equally often during non-REM sleep, the normal overnight learning was found the next morning.

Sports managers and trainers are very concerned that their athletes get proper sleep in the weeks and months leading up to competitions. It seems likely that the skills of movement and perception they have honed during each day's practice may be consolidated during REM sleep.

Rats—like most other animals—also dream. When rats run through mazes, the brain cells in the memory center of their brains—the hippocampus—fire in a particular pattern that is different for every maze. Each maze, then, has an electrical "signature" in the rat's brain, and scientists have searched for this signature in a rat's REM dreams. Sure enough, they have identified rats dreaming about particular mazes, and with an accuracy that let the researchers know which particular point in the dream maze the rat had reached at each moment in the dream. These rat dreams weren't necessarily connected to the experiences they had just had; the REM dreams may have been linked to more distant experiences. Non-REM sleep—in the case of you or I—is perhaps more likely to be the venue for gnawing over the past day and its events and worries.

Where Dreams and Waking Meet

Dreaming is an alteration of our normal daytime state of consciousness and depends on some parts of our brain becoming less active while others increase their activity. REM dreams are overwhelmingly about imagery—in the widest sense of the word in all sense modalities—but it is imagery without the checks of the frontal lobes and their management role in the brain.

There are some states in between waking and sleep, however, where the heedless drama queens of REM dreamland find themselves at the same cocktail party as the sober prefects of the prefrontal cortex. You may well have found yourself half awake and half asleep either at bedtime or in the morning and drowsily meandered through this no-man's land of consciousness—the so-called hypnogogic and hypnopompic states. In these states of mind, you can both be aware that you are dreaming and be dreaming at the same time—seeing, feeling, and doing in the mind's eye, but half-realizing that this is dream and not reality. In these twilight zones, the frontal lobes have not fully closed down, yet the brain chemistry has changed to alter the activity in the brain's imagery regions and in its memory centers.

A rarer but more dramatic example of this is *lucid dreaming,* where you are fully in a REM dream state but have the awareness that you are dreaming at the same time. Again, a likely part of the explanation for this is that the frontal lobes are active enough to let you be aware of what your brain is up to, but not active enough to suppress all the other apparatus associated with the dream state.

Some people who experience mystical or out-of-body states may in fact have slipped into lucid dreaming, and it also *may* be possible to train yourself to dream more lucidly—to be more aware of the fact that you are dreaming when you are dreaming. In one study, researchers at McGill University in Montreal found that recurrent nightmares suffered by five adults were virtually eliminated by training them to use lucid dreaming during the nightmares to change them into less frightening events. A key part of the training was getting the individuals to use the mind's eye to visualize the nightmares and change the images linked to them. With practice, it seems, they managed gradually to transfer this skill into the dream state itself, and could dream lucidly at least some of the time.

Such training of the mind to change its state is central to some religious and quasi-religious practices. Sleep and dreams are a natural way of altering consciousness and distorting time, space, and self. Drugs are another way in which this can be achieved, and as you saw earlier, hypnosis and mental imagery are also techniques that can change consciousness in quite dramatic ways. Meditation is a method of altering consciousness, one that has arisen, most familiarly, out of Buddhist and Hindu traditions, though most religions have some degree of contemplation, focusing of consciousness, or ritual comparable to meditation. In some type of meditation, practitioners are taught to expect alterations in their sleep and dream consciousness akin to those of lucid dreaming.

These ancient methods seek to foster a type of mute awareness during sleep—which they often seek also to extend into everyday waking life. This "silent watcher" seems to have similarities to the toned-down prefect of the frontal lobes during lucid dreaming or hypnogogic trances: watching and aware, yet detached and not trying to change things. One study of very experienced meditators found that those who said they had reached the stage of having this sleep awareness—through many years of meditation practice—did indeed show distinct differences in electrical brain activity during sleep. Specifically, the researchers found that a brain wave pattern typical of deep sleep coexisted with a brain wave pattern associated with relaxed alertness, while the subjects were apparently fast asleep.

It seems possible, then, that we can alter the many and varied states of consciousness that exist between coma and total alertness through various methods, ranging from hypnosis to meditation. Mental images are central to most of these methods, and religions throughout history have drawn on both the amazing human capacity for mental imagery and our remarkable ability to alter our state of consciousness at will. This is the subject I turn to in the final chapter.

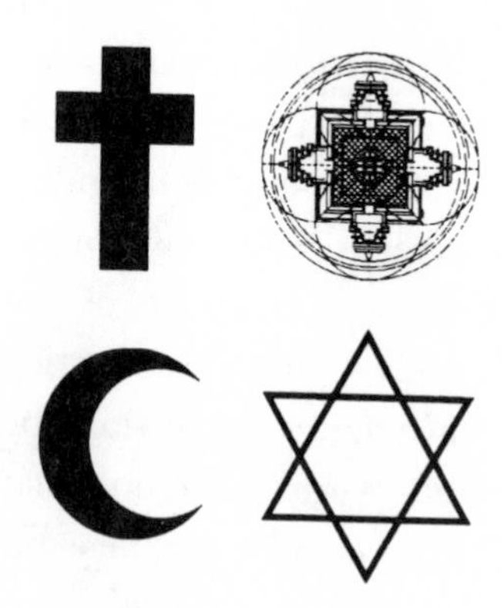

12

Images of God

It is October, 312 A.D. The would-be Roman emperor Constantine and his troops are marching toward Rome to do battle with his opponent, another would-be emperor—Maxentius. Constantine stops to pray, trying to combat Maxentius's magical enchantments. As he prays, just after noon, he has a vision. He sees the sign of the cross emblazoned across the sky. Constantine goes on to win the battle of Milvian Bridge and thereafter has this vision—the cross—inscribed on the armor of all his armies. With this man's vision, world history changed, the world's greatest empire espousing a new religion. From this flowed the most enormous political, social, and physical changes in the world, the effects of which are still with us to a fundamental extent today.

Visions and their images are the marrow within the bones of human meaning. This is so in religion, in politics, in sports, and in war. Think of the Nazi swastika, the image of the meditating Buddha, the yin-yang symbol, the Star of David, the cross, the hammer and sickle, the rising sun. In business too, images rule, with large corporations spending millions of dollars to find the right brand icon, from the McDonald's golden arches to the Nike swoosh.

Many, many religions have been based on the visions of their founders, powerful mental images that in one stroke captured spectra of meaning and nuances of emotion transmittable like lightning to other human minds. Bypassing the brain's language circuits, these images can be communicated across cultures otherwise divided by language.

Because images are outside language, they also can free the mind from the constraints of logic. They can change how we behave by the emotions they can so readily arouse and the unverbalizable meanings they evoke. Men, women, and children have been slaughtered, wars waged, and whole cities razed over whether these visions had real correlates in the physical world or existed nowhere but in the brains of the visionaries.

The mind's eye is so powerful that it can create mental worlds so vivid as to be indistinguishable from the external world. In fact, it is a miracle that we can—mostly—tell the mental image's firing of brain cells from the activity of near-identical sets of brain cells registering the outside world. Sometimes we can't—in dreaming, for instance, or under hypnosis.

So does that mean that all religious visions can be reduced to the excited fluttering of a few million cells in the brains of visionaries? One could answer yes to this question, but it would be as facile as saying that Michelangelo's creative vision, which led him to sculpt the statue of David, is reducible to such neuronal activity. The subtle beauty of the work of art depended *on the firing of Michelangelo's brain cells but is not* reducible *to this. To try to understand the artistic vision and its impact, we have to draw on realms of aesthetics and human emotion that cannot simply be written off as meaning-free brain activity.*

And so it is for religious images. Packed as they are with subtle symbolism and hard-to-verbalize meanings, their content cannot be written off as "mere imaginings." As the famous ballerina once said when asked to explain the meaning of her dance: "If I could explain the meaning, I wouldn't have to dance it." In dancing the unverbalizable meaning of her dance, the ballerina was conveying meaning through that great sea-channel of wordless consciousness that we tend to neglect in our everyday life. The images of eye and body that shaped the dance not only cannot be explained by words—they would most probably be strangled by them.

Of course, many visionaries have been psychotic, plagued by hallucinations that they interpreted as messages from God. The uncontrolled firing of brain cells in epileptic seizures have also triggered hallucinations and emotions that have been taken as relevatory messages. But artists such as Vincent van Gogh have produced great art that probably depended in part on abnormalities of brain-cell firing, yet that does not devalue their art.

As with art, it is not the role of brain scientists to judge the worth and validity of religious experiences and images. The fact that a particular religious image can be analyzed in terms of visual imagery based on firing of brain cells in the left occipitoparietal cortex says nothing about the wider meanings and symbolism of that image within its host theology.

As a scientist myself, I come with this modest perspective to the question of religious practices and beliefs, a realm of human experience where the mental image often reigns supreme.

God in the Brain

Religions bind human minds through the power of imagery. The central images of each religion—usually visual or auditory—are similarly evoked in the minds of millions during ritual and devotion, and can transcend culture, race, and language. These common images may help break the boundaries of individual consciousness, creating common meanings and purpose that make up the glue of social cohesion. Because images are wordless, they transcend logic, and the wordlessness of their meanings makes sensible dialogue between religions very difficult. Hence the downside of religion—nonrationality and the intolerance and strife that can go with it. Yet secular societies that have lost the common binding images of mass religion suffer because their citizens are more likely to feel isolated and alienated: we know, for instance, that countries with higher levels of religiosity and religious beliefs tend to have lower suicide rates—among men, that is.

But imagery in religion is more than just a social glue binding people together with common pictures in their mind's eyes. The rituals, prayer, and meditation of most religions help release people from the "cool web of language" and allow them access to the world of images. Many of us have little opportunity for escaping out of the world of words and into the world of image, except in our dreams—and possibly also in music or art. As mass religions decay in many Western countries, many of us are turning to practices such as meditation derived from Eastern religions like Buddhism. More mental techniques than religious faith for many practitioners, they nevertheless allow people access to the word-free realms of consciousness. Using these methods, people can achieve states of mind and qualities of consciousness that simply aren't on offer in most of everyday life.

At least—that's what they tell us. How do we know they aren't just making it up? Fortunately, brain imaging lets us study in detail what's actually going on in people's brains when they meditate. And what we find is that the changes in the brain correspond quite closely to the experiences that people say they have. Let's take one common experience that people have during meditation, prayer, and other religious practices. This is the feeling of one's body dissolving into the surrounding space—part of a sense of oneness with the world around, and a breaking down of one aspect of the self—the bodily self.

When meditators who achieve this experience were studied using PET brain scanning, the researchers found that the brain activity of the parietal lobe of the right half of the brain had decreased. It is in this part of the brain that the body map is stored; people with damage to this part of the brain can have strange distortions in how their body feels—even to the point of feeling that they have extra limbs, or misshapen body parts.

Certain types of meditation, then, seem to be able to change the way your body feels to the point that you temporarily lose a sense of the boundary between your body and the outside world. This sensation is "real" insofar as the brain region that stores the information about the body boundaries is—so to speak—put to sleep by the meditation.

Sometimes people who are engaged in meditation, prayer, or other religious rituals also feel as if time is becoming distorted, and some even report that they enter a sense of "timelessness." Just as the feeling of the shape of our bodies is stored by a particular pattern of firing in certain neurons, so the sense of time, and our progression through time, is stored by different patterns of firing in different sets of neurons. If the meditation or prayer dampens down or changes the patterns of firing of these neurons, then the sense of time can be distorted, through brain changes that correspond very well to changes in what these people experience.

One perspective on this might be, "So what? These people are just tricking themselves into imagining reality as being other than it is—it's all self-delusion and playing games with the brain, just as you do if you take drugs." This is a fair point, but falls into the trap of assuming that what the brain normally represents is "reality." Take space and time, for instance. We know from Einstein—the great visualizer—that a continuum of space and time exists, that time-space is curved, and that time changes the faster we travel through space. Yet none of these facts about reality are coded by our brains in our everyday life. We do not *experience* time-space as curved, and we don't *experience* time changing according to the laws of physics. We don't experience these things because the dimensions of space and time within which our brains have evolved, and within which we grow up, are too narrow to make the action of these laws apparent. In other words, the human world is too small and slow for the action of these grand laws pertaining to unimaginable distances and speeds to be relevant or measurable.

Our brains have evolved and learned, then, to code a tiny range of the infinite spectrum of time and space. When the meditator then experiences "timelessness," it is no more of an illusion than the sense of "real" time spinning by in the head of the busy commuter. The sense of time passing is as much an arbitrary construction of our brains as the sense of space is. When, during meditation, prayer, or other ritual, a person experiences an altered sense of time and space, this is in many ways just as "real" as the normal everyday experience of space and time.

Meditation and other altered states of consciousness therefore yield alternative subjective realities that may be as valid as the "normal" subjective realities. Imagery seems to be very important in many such altered states of consciousness, depending on the precise type of meditation used.

There are hundreds of different types of yoga practice, one of which is called Yoga Nidra. In Denmark, researchers studied nine experienced teachers of Yoga Nidra as they meditated. They found that the parts of the brain that we know are involved in imagery became active during meditation, whereas the regions in the frontal lobes of the brain that allow us to make plans, contemplate the future, and worry were dampened down.

Who Am I?

As I sat talking to the woman, she looked down at her right hand with fear in her eyes. "Oh, God, it's starting again," she said. She leaned back as the clawed hand strained toward her throat. She tried to fight it off with her left hand, but it was stronger, and it caught hold of her collar, twisting the cloth on her neck. With a huge effort, the woman managed to wrestle her alien hand back to the table.

This woman's stroke in the left frontal lobe of her brain had also cut the connections between the left and right front halves of the brain. This meant that any actions set off by the left half of the brain couldn't be inhibited by the right frontal lobe. As I could see from the fear in her eyes, to the woman this hand was alien—it wasn't "hers," it didn't have the sense of "me-ness" that our actions usually have. In many parts of the world, this unfortunate woman would be described as "bewitched" or taken over by some external force or spirit. Three hundred years ago she might have been burned as a witch—maybe much more recently in some places.

Yet this alien hand was simply a result of damage to the brain causing a disruption in the transfer of information between the two sister halves. The right half of the brain seems to be particularly important in imparting that sense of "me-ness" to our actions and memories. So when the woman's right hand rose to her throat—controlled by the left half of the brain, of course—the isolated right frontal lobe wasn't able to impart this sense of "me" to the action because of the broken connections with the left frontal lobe where the action was controlled.

In Chapter 6 I showed you how the "machinery of self" can break down in memory too. The right hemisphere of the brain seems to be the seat of a self-knowing awareness that gives us the thread of continuity back into our pasts and forward into the future. It allows us to have the sense of being a continuous entity over time, and lets us feel that what our hands are doing right now is caused by us and not by some alien external force.

His left arm lay limp by his side. "Can you move it at all?" I asked him. "It's not mine," he replied. "Not yours?" "No, it's attached to me, but it's not mine." "Where did it come from?" "I've no idea, but I hate it, and I want it taken off."

This was another of my patients, who had had a large stroke in the right half of his brain. You could have a perfectly rational, sensible conversation with him about anything except his left arm. That paralyzed arm felt completely alien to him—not in a metaphorical sense but in a real, physical way. He genuinely believed that it was not his arm. The arm was paralyzed because of the stroke, and he had no feeling in it, so there was no sensory information coming from the arm except through his eyes. He could *see* his arm, but there was no sense of "meness" about it.

I have seen very many patients who have lost the sense of coherent bodily self in this way. The vast majority had damage to the right half of their brains—lending support to the view that the right half of the brain is very important in giving what we do, think, feel, and remember that sense of "me-ness."

"How are things, Chris?" "Okay, I suppose, for someone who's dead." "I beg your pardon?" "I'm doin' okay for someone who's dead." "You're dead?" "Yes." "You mean you feel like you're dead?" "No, I'm dead. I am not living. I am dead."

This young man—articulate, completely normal in appearance, witty, and charming—had crashed his motorbike and damaged his brain, particularly on the right side. So completely had he lost the sense of "me-ness" that *nothing* he felt, thought, did, or remembered felt to him as if he was the person experiencing it. Rather logically, then, he concluded—no *felt*—that he had died and this was the afterlife.

The "I" that is reading this page is a rather precarious construction, then. It depends on the correct firing of millions of neurons, particularly in the right frontal lobe of your brain. Some of you who have been in a severe life-threatening situation might have had the experience of watching yourself in a detached way as if it were not you who was in danger but some other person identical to yourself.

Severe stress or danger can give such out-of-body experiences, as can hypnosis, borderline states between sleep and waking, epileptic-type activity in the brain, and hallucinogenic drugs such as LSD. Like visions, these distortions of self can be interpreted by some people as religious messages from an external force, rather than as understandable fluctuations in brain function due to temporary or permanent disturbance to the activity of neurons. Roughly 70 percent of us have at some time an altered sense of self that we interpret as a spiritual or religious experience. Maybe most of the blockbuster religious visionaries throughout history could have their visions explained by modern neuroscience?

Perhaps they could, but as with space-time perception, our sense of self is a rather arbitrary and provisional construction, as the cases I described above show. When people experience during meditation or prayer a sense of their self changing, blending, and unifying with a wider reality, this is no more illusory than the normal sense of self. Indeed one could make a reasonable argument that is more like "reality" than is the normal sense of self.

Language and the left hemisphere tend to dampen down imagery and other wordless experiences, including some experiences that are commonly interpreted as "spiritual." So some types of prayer, if overwordy, might reduce the opportunity for the experiences that ritual, chants, and nonverbal meditation or prayer can evoke. The left hemisphere also has a much tighter system of semantics—meanings and concepts. Meanings and the logic of concepts are beautifully engraved into the trembling web of interconnected neurons in the left temporal lobe of the brain. It is here we can distinguish the concept of "guilt" from that of "shame," of "ocean" from "sea," of "necessary" from "sufficient," and so on.

In the right hemisphere, there is a less precise semantic system where the same concepts—heavily image-linked—are more loosely connected with each other, more overlapping, and less constrained. Someone relying on the right hemisphere semantic system for reading (say their left hemisphere language system is malfunctioning) might read the word *table* but say the word *chair,* for instance.

You saw in Chapter 5 how creativity often depends on these looser connections, and how imagery can act as the essential "loosener" of tightly woven concepts, allowing more creative and original thinking. The "cool web" of left hemisphere language tends to inhibit this type of thought through its beautifully crafted semantic systems. When such thought is disrupted—by brain damage, for instance—people can become prone to supernatural experiences, such as seeing great meaning and import in purely random events. Coincidences that the conceptually tight left hemisphere would see as statistically random events are more likely to be viewed by the conceptually looser right hemisphere semantic systems as being meaningfully connected in some way.

People who have supernatural beliefs—who for instance believe in ghosts, precognition, and other ESP phenomena—tend to make more use of the loose and creative right hemisphere semantic systems. Take this example from the work of Peter Brugger, a neurologist in Zurich. Supposing two people—one prone to belief in supernatural, magical processes and the other not—had a dream about winning a car. The next day, they both have their bicycles stolen. The second, nonmagical thinker would, if she thought about the dream at all, write the two events off as a coincidence. She would probably be relying more on left hemisphere semantic systems where the concept "car" activated a precise and circumscribed set of neurons, overlapping only slightly with the set of neurons activated by the concept "bicycle." The same would be true for the concepts "stolen" and "won."

The magical thinker, on the other hand—presumably relying more on the looser semantic systems of the right hemisphere—would have a much greater overlap in the neurons activated by these concepts. "Car" and "bicycle" would each activate a much broader and more overlapping group of neurons, and the same would be true of "won" and "stolen." Because of this big overlap, the magical thinker would understandably see a much greater degree of coincidence in the dream and the event than the nonmagical thinker. And because of this, he would be much less easily convinced that the two events could have happened by chance—by mere coincidence. He may well see precognition here, and turn to other explanations, such as the prophetic power of dreams.

Peter Brugger, one of the world's leading researchers on superstition and magical thinking, has assembled a lot of evidence to suggest that they are related to an increased reliance on the looser and imagery-related networks of the right hemisphere. The same things that make these people superstitious also tend to make them more creative than average, however. So it isn't just when the brain is damaged that you find people experiencing outside forces influencing their behavior. Normal variations in the way brains work also make people more or less susceptible to seeing forces at work in events that may in fact be statistically random.

While language and its logical pathways can inhibit imagery and all its benefits, so overreliance on imagery can corrode logic and all *its* benefits. But both types of thinking are limited, and both are equally necessary for balance. God help the organization or country whose leaders are completely dominated by either of these modes of thought.

Magical thinking is probably a key cause of failure of economic and social development in some of the world's poorest countries, going hand in hand with poor education. Education, of course, largely focuses on the "cool web of language," and as you saw in Chapter 2, this focus leads to a snuffing out in our children of their capacity to use the nonverbal capacities of the brain that are the subject of this book.

Perhaps, then, we need to consider a parallel series of problems to those caused by superstition and a neglect of the language-based systems of the brain. Perhaps we need to consider the potential loss of things such as creativity and health-improving and memory-improving imagery caused by the neglect of the mind's eye. Maybe the parallel to the economic chaos and poverty of countries that rely on magical thinking is the lack of creativity and sensitivity of developed countries in dealing with the problems of development and the environment?

A Mindful Future

By 2020, depression will be the second most common disease in the developed world. Suicide rates throughout the developed world are rocketing, particularly among young men. What is happening to us? I am not a sociologist, and there is no single, simple answer to this question.

Rapid social and economic change seems to be a factor, however. Perhaps people—particularly young men with their macho emotional inarticulacy—find it hard to create meaning in the world as the world changes around them. Our thoughts, our emotions—our very sense of who we are as individuals—have to be constantly recalibrated and reevaluated as the world changes. If we don't have the means for such recalibration, we may give up in despair. But what do we need to keep the changes in our self-identity up-to-date with the changes in the world? Language and logic, of course, but these may not be enough. We probably also need the capacity for examining and changing our emotional responses, as well as for being able to modify our experiences of self. Most people—particularly young men—are not well equipped to do so.

I have shown you how important our nonverbal brain systems are in controlling stress, in improving health, and in fostering a whole set of experiences that our much-needed "cool web of language" is less efficient in dealing with. Yet most countries have virtually no ways of educating these brain systems. In contrast, they are pretty good at tutoring the "cool web of language" in mainstream education. Some religious practices seem to stimulate these nonverbal brain systems, and recently the secular scientific world has come together with its religious counterpart in producing some very exciting developments that might help us come to grips with the need to foster the nonverbal as well as the verbal capacities of the human brain.

Mindfulness meditation derives from a branch of Buddhism and trains people to develop an "aware" mode of consciousness that contrasts with a mode dominated by habitual, goal-oriented habits of thought and feeling. Mindfulness meditation teaches people to focus their awareness on what they are experiencing in the moment, including thoughts and emotions as they run through their minds. They further learn to be open and accepting of the content of these thoughts and emotions, and to be aware of them with a certain observer's detachment. Mindfulness meditation is a way of taking at least some of the "me-ness" out of awareness, of becoming, to some extent at least, a partly detached observer of the billions of thoughts and feelings that course through the mind every hour.

In some ways, this type of "observer status" for the self better reflects the reality of what is going on in the brain than the "field commander" status of our normal day-to-day consciousness. As you saw earlier with my clinical cases, "me" is a very fragile and multilayered construction, maybe even a somewhat arbitrary one. "Me" consists of feelings, thoughts, memories, plans, intentions, frustrations, worries, and all the rest. "I" am—most of the day at least—the battlefield commander, immersed in the struggle that is mental life, making and achieving goals—or not, pitched up and down on the emotional treadmill, worrying about the future, reflecting on the past, and so on.

John Teasdale of the Cognition and Brain Sciences Unit in Cambridge, England, is one of the world's leading researchers on depression. With colleagues in Canada, he integrated the Buddhist method of mindfulness meditation with Western methods for treating depression, calling this new way of mind-training "mindfulness-based cognitive therapy."

Cognitive therapy is a way of helping people to control mood and improve depression through modifying negative thoughts linked to low mood. This mainly language-based therapy has been very successful in helping combat depression and prevent relapse in people who have been depressed. Teasdale's approach combined this verbally based therapy with methods derived from mindfulness meditation developed by Jon Kabat-Zinn of the University of Massachusetts. Rather than focusing on *what* people are thinking, Teasdale's approach is to train people to alter their *awareness of* and *relationship to* their thoughts and feelings.

In standard cognitive therapy, for instance, people will be trained to monitor their own thoughts and catch themselves in the act of thinking thoughts like "I am a failure," "People think I'm stupid," or "I'll never meet anyone who can love me." In depressed people these thoughts can become habits—and like most habits can become so automatic that you are hardly aware of them. Cognitive therapy is a powerful, language-based training system that gradually trains people to become aware of what negative thoughts they are thinking and to notice how these bring down their mood. Based on the assumption that these thoughts are usually inaccurate and overly negative, people are trained gradually to replace the automatic negative thoughts with more positive ones that won't bring down their mood.

John Teasdale's group further developed this successful method by capitalizing on the mind's nonverbal capacities for changed awareness and experience. They trained people who had suffered serious clinical depression, but who had recovered for the time being, in mindfulness-based cognitive therapy. This training loosens the link between "me" and my thoughts, feelings, and bodily sensations.

On one hand, the "me" battlefield commander's thought "My God, I'm going to lose" will be fully and completely "mine." Such an emotion-laden thought, and all the sensations of bodily dread and anxiety that go with it, will fully occupy "me," so that "me" and the thoughts and emotions are more or less one. The mindfulness training, on the other hand, fosters a kind of UNobserver type of awareness in the conscious mind. Rather than being sucked into the emotions, the anticipated consequences, the ramifications, the anxious planning that a thought like "My God, I'm going to lose" will usually trigger, people are taught to be aware of the thoughts and feelings running through their minds but to be aware of them in a detached, observer-like way.

If I am totally immersed in the "cool web" way of thinking, "I" and my thoughts are one. If the thought that I am going to lose the battle comes to dominate the "cool web," then for as long as that thought dominates, that is "me," with all the negative emotions and further thoughts that this conclusion will generate. Conversely, if I have the triumphant thought "I'm going to win," then that thought is "me," as are all the exuberant emotions and optimistic thoughts and plans that will follow. But as you have seen in this book, our consciousness isn't really a single river flowing to the sea. At any one second your brain is processing multiple and parallel computations with billions of pieces of information, only a tiny fraction of which reach your awareness. Consciousness is more like a river delta with multiple small rivers spreading out toward the sea. Language is only one section of this river delta.

How much of the river delta of consciousness we actually become aware of from moment to moment depends on how and what we attend to. If we attend only to the language section of the delta and its wordy thoughts, then these will fill our consciousness and we will be aware of little else. Similarly, if we only focus on the images and sensations of other parts of the delta, we may be sucked into a roller coaster of emotion that can come to dominate our consciousness.

If, on the other hand, we zoom back to an aerial view of the delta, then both the verbal and nonverbal parts of the river system can be brought simultaneously into consciousness. When we become aware of these multiple rivulets in our consciousness, some of them contradicting each other, it is harder for "me" to become too caught up in any one stream of thought or feeling. The UNobserver "me" can see pros and cons in the various sides and maintain a certain sympathetic neutrality.

Unlike the battlefield commander, the UNobserver sees the thought "I am losing," the image of defeat, or the bodily sensation of dread as just what they are—somewhat arbitrary thoughts, images, and bodily sensations that arise because none of them can see the whole delta. The observer sees them for what they are—very limited perspectives across the riverbanks.

Mindfulness training taught the people in Teasdale's study to learn to take this sympathetic but detached view of their own thoughts, images, and feelings, not allowing "me" to become caught up in the current in any one of the rivers, and maintaining a balanced awareness across the delta of conscious experience. They achieved this through learning from time to time to suspend the normal style of day-to-day thinking and change their brain state with this particular type of meditation. And the people who had relapsed into major depression at least three times in their lives before had a significantly lower chance of relapsing for the year after the mindfulness training, compared with the control group who got normal treatment for their depression.

The Long Search

God is dead, Nietzsche famously told us. As formal religion collapses in Western Europe and many other parts of the developed world, new beliefs and practices expand to fill the vacuum. Scientists who oppose religion wring their hands in despair—how can apparently intelligent people believe impossible things? Ghosts and spirits, energies and telepathy, astrology and hexes, angels and archangels—how is it that a rational, educated society can sustain such antiscientific vapors of the mind? these scientists plaintively ask.

It is certainly true that superstition and magical thinking are like glue in the wheels of enlightened progress in underdeveloped countries. Without the sweet reason that grew out of the Renaissance we would have no scientific progress, no great cities, no space exploration, no average life expectancy into our seventies and eighties. Without the "cool web of language" and its sister sweet reason, our children would be dying like flies of smallpox, diphtheria, typhoid, and a thousand other killer diseases. Scientists like Louis Pasteur would not have uncovered the causes of such diseases had they been addicted to loose, magical thinking that tolerates internal contradiction and illogic.

Science is the child of logic, and antiscientific superstition is a major threat to civilization. Science has, however, produced costs as well as benefits to mankind: weapons of mass destruction, global warming, and environmental pollution, for instance. Religion, on the other hand, is very much a child of the nonlogical domains of the brain, with visions and imagery particularly playing a big part in religious revelation.

Language and logic represent only one way in which our brains can make sense of the world. In this book I've shown you how there are other ways in which our brains capture and represent the world. These wordless domains have been neglected in Western civilization, particularly in the twentieth century. These domains defy logic—reason is simply not their bag. But what they lack in cool rationality, they make up for in cinematic vividness, emotion-rousing pungency, and consciousness-changing potency. Imagery burns through to the very membranes of our cells, changing our physiology and changing our immunity for better or worse. Creativity flourishes in these domains of mental life—but the very same processes that foster creativity also make us vulnerable to superstition and magical thinking.

Think of the people you know—friends, family, and workmates. Some will be cool, clear thinkers, doyens of the "cool web" of logic and language, while others will be mentally loose-limbed, intuitive, superstitious, not overprone to logic, but probably quite creative. Most will probably be somewhere in the middle, switching between the two modes—reading the science pages in the newspapers most days, but guiltily glancing at the astrology column in the tabloid on a Sunday morning.

Where do you fall in this spectrum? It takes all sorts to make a world, but it seems to me that to devote oneself entirely to just one of these mental modes is the equivalent of driving a convertible but never putting the roof down. Great scientists like Einstein need the roof down at times to help them think creatively and intuitively, and great artists, visionaries, and spiritual leaders similarly must—sometimes at least—put up the roof of logic on their sometimes wacky visions.

Religion can be roof-down or roof-up. Words dominate some—take Christian Protestantism, for instance—while images dominate others—Hinduism, for example. Devotional practices can be highly verbal—Christian prayer, for instance—or they can be based on inducing altered states of consciousness through mental imagery, meditation, music, ritual, chants, dance, and many others.

The human race will never be completely logical, no matter how fine its education systems, for the simple reason that the human brain works in nonlogical as well as in logical ways. But neither should the human race ever be entirely logical, because that would mean welding shut the hood of the convertible and cutting ourselves off from an immensely rich domain of unlogical experience.

As Stephen Jay Gould argues in his book *Rocks of Ages,* the scientific-logical and religious domains do not overlap. Each has its own realm of learning. The problems arise between science and religion when one tries to interfere with the other. So we had Galileo persecuted by the Church for heliocentric heresies on the one hand, and religion denigrated by antireligious scientists such as Richard Dawkins for its intercourse with the nonlogical domains of human experience on the other.

Of course, theology runs into problems when it tries to drive with the hood up. Logic can only partially and imperfectly encompass the nonlogical realms of experience. Do I, for instance, as a scientist believe in reincarnation? No. Do I believe in the resurrection of the body? No, I don't. But as a vaguely yearning agnostic I can see the virtue in intermittently and judiciously suspending scientific logic to enter the "as if" of religious drama. Why? Because I believe that most religions have evolved ways of leading their adherents to realms of experience that are otherwise shut off to them. True, many—indeed most—people have experiences they describe as transcendent outside of formal religious settings, but the cultivation of sustained altered states of consciousness has largely been the role of various religions, albeit with varying degrees of success.

These realms of experience lie, I believe, largely outside the realm of the "cool web of language." Sophisticated theologies use reason and logic to provide a structure within which these experiences are ordered and explained, and that structure is usually internally quite coherent. Within the borders of the theology, this coherence is often logical, and religious images often have a sophisticated foundation of rationally explicable symbolism. They are usually also tied in with a logically coherent system of morality, and this whole edifice of theology is a domain of human experience codified in a way similar to legal systems.

But at the heart of most religions are the visions and wordless experiences that the human brain yields to us. We don't need religion to have these experiences—but most religions have, somewhere at their core, the wordless visions of some men or women, somewhere. And religions have developed ways of cultivating these visions and experiences, just as secular culture has evolved educational systems to cultivate the "cool web of language" and logic.

The more we understand about the human brain, the more our education systems must change to better cultivate the powers of both language-logic and imagery-intuition. And so it is for religions too: religious beliefs, like scientific hypotheses, are tools—a means to an end—and neither can ever be absolute and final truths because of the severe limitations of the human brain for understanding the world and ultimate reality.

Perhaps religion has further to go than science in recognizing this fact. While some scientists lambast religion, they only get away with it because many religious figures continue to use theological arguments in the realm of science, where they really have no business. Perhaps religions will evolve in this century into what—and this is a personal opinion—should be their proper role: as custodians and tutors of the realms of transcendent experiences and morality. This will mean yielding up to their sibling science the task of sketching the nature of objective physical reality.

If this happens, then scientists will no longer be justified in denigrating religion, and religion will give up its unequal struggle against scientific explanations of the physical world. Such a synthesis will be one step toward an overdue and equal cultivation of the two great realms of the human mind—language and imagery.

Notes

Chapter 2

p. 10 . . . *weren't really "art."* Jaynes, J. (1979) Commentary on Haber *et al.*, Eidetic Imagery: *The Behavioural and Brain Sciences* 2, 605–7.

p. 11 . . . *Nadia drew this.* Selfe, L. (1978) *Nadia: A Case of Extraordinary Drawing Ability in an Autistic Child.* New York: Academic Press.

p. 15 . . . *different-looking person.* Simons, D. J., and Levin, D. T. (1998) Failure to detect changes to people during real-world interaction. *Psychonomic Bulletin and Review* 4, 644–54.

p. 20 . . . *The Mind of a Mnemonist.* Luria, A. R. (1968) *The Mind of a Mnemonist.* Cambridge: Harvard University Press.

p. 22 . . . *England, proved this.* Hitch, G. J. *et al.* (1988) Visual working memory in young children. *Memory and Cognition* 16, 120–32.

p. 25 . . . *five to six is ten.* Waldfogel, S. (1948) The frequency and affective character of childhood memories. *Psychological Monograph* 62, Whole Number 291.

p. 26 . . . *eight months later.* Bauer, P. J. (1996) What do infants recall of their lives? *American Psychologist* 51, 29–41.

p. 28 . . . *event, and how.* Pillemer, D. B. *et al.* (1994) Very long-term memories of a salient preschool event. *Applied Cognitive Psychology* 8, 95–106.

p. 29 . . . *one crucial ingredient is required*. Perner, J., and Ruffman, T. (1995) Episodic memory and autonoetic consciousness: Developmental evidence and a theory of childhood amnesia. *Journal of Experimental Child Psychology* 59, 516–48.

p. 31 . . . *had imagined it*. Newcombe, N. S., *et al*. (2000) Remembering early childhood: How much, how and why (or why not). *Psychological Science* 9, 55–8.

Chapter 3

p. 35 . . . *most famous neuropsychological studies ever*. This is an imaginative reconstruction of the interviews carried out by Professor Bisiach and Dr. Luzzatti in their famous study of mental imagery in patients suffering from unilateral neglect. Bisiach, E., and Luzzatti, C. (1978) Unilateral neglect of representional space. *Cortex* 14, 129–33.

p. 38 . . . *experience in words*. Professor Michael Gazzaniga of Dartmouth College has pioneered research into split-brain patients. A description of these and other studies can be read in various books and articles, including the following: M. Gazzaniga (1998) The split brain revisited. *Scientific American* 279, 50–5.

p. 39 . . . *in real life*. Shepard, R., and Meltzer (1971) Mental rotation of three-dimensional objects. *Science* 171, 701–3.

p. 39 . . . *this may happen*. Georgeopolis, A. P. *et al*. (1989) *Science* 243, 234–6.

p. 40 . . . *direction of the neurons*. As single-cell recording cannot be done in humans, we cannot be sure yet that mental rotation of the type you did in the exercise follows exactly this pattern. Recent functional imaging studies of the brain do, however, give indirect support to this notion.

p. 42 . . . *mental map reading*. Farrell, M., and Robertson, I. H. (1998) Mental rotation and the automatic updating of body-centered spatial relationships. *Journal of Experimental Psychology: Learning, Memory and Cognition* 24, 227–33.

p. 44 . . . *of their brains*. Farrell, M., and Robertson, I. H. (2000) Automatic updating of egocentric spatial relationships: Impairment due to right posterior lesions. *Neuropsychologia* 38, 585–95.

p. 45 . . . *type of damage*. Farah, M. J. (1991) *Visual Agnosia*. Cambridge: MIT Press.

p. 46 . . . *of your head*. Ungerleider, L. G., and Mishkin, M. (1994) What and where in the human brain. *Current Opinion in Neurobiology* 4, 157–65.

p. 47 . . . *were most active*. Haxby, J. V., *et al.* (1994) *Journal of Neuroscience* 14, 6336–53.

p. 47 . . . *King Duncan's blood. Macbeth,* Act II, Scene I by William Shakespeare.

p. 47–48 . . . *into visual imagery*. Stephen Kosslyn has written two fine books about mental imagery, from which many of these examples are taken. These are *Image and Mind,* 1980, Harvard University Press, and *Image and Brain,* 1996, MIT Press.

p. 48 . . . *above your ears*. McGuire, P., *et al.* (1996) Functional anatomy of inner speech and auditory verbal imagery. *Psychological Medicine* 26, 29–38.

p. 50 . . . *obscure your vision*. Stephen Kosslyn, *Image and Mind,* 1980, Harvard University Press, and *Image and Brain,* 1996, MIT Press.

Chapter 4

p. 51–52 . . . *lead me astray*. This case by Charcot, including a translation of the letter from M. X, is described by Professor Andy Young of the University of York and Dr. Claudia van de Wal of the University of Leiden in the Netherlands, in the following text: Young, A. W., and van de Wal, C. (1996) Charcot's case of impaired imagery. In C. Code, C-W. Wallesch, Y. Joanette, and A. R. Lecours (eds.) *Classic Cases in Neuropsychology*. Hove, UK: Psychology Press, 31–44.

p. 53 . . . *their mind's eye*. This questionnaire is reproduced with the kind permission of Professor David Marks and the British Psychological Society, from Marks, D. J. (1999) Consciousness, mental imagery and action. *British Journal of Psychology* 90, 567–85.

p. 55 . . . *your visualization ability is*. These approximate figures are based on a comprehensive review of dozens of studies carried out in the following paper: McKelvie, S. J. (1995) The VVIQ as a psychometric test of individual differences in visual imagery vividness: A critical quantitative review and plea for direction. *Journal of Mental Imagery* 19, 1–65.

p. 56 . . . *walking toward you*. See McKelvie (1995) op cit.

p. 58 . . . *visual imagery*. Sheehan, P. W. A shortened form of Bett's Questionnaire upon mental imagery. *Journal of Clinical Psychology* 23, 386–9.

p. 60 . . . *as a rabbit*. Picture from Jastrow, J. (1990) *Fact and Fable in Psychology*. Boston: Houghton Mifflin.

p. 60 . . . *Trieste, Italy, asked*. Bradimonte, M. A., and Gerbino, W. (1993) Mental image reversal and verbal recoding: When ducks become rabbits. *Memory and Cognition* 21, 23–33.

p. 62 . . . *"Monday occurs about four times per month."* Eddy, J. K., and Glass, A. L. (1981) Reading and listening to high- and low-imagery sentences. *Journal of Verbal Learning and Behaviour* 20, 333–45.

p. 64 . . . *deliberately switched on*. For an early but excellent review of this and related research, see the following article by the eminent neuropsychologist Martha Farah. Farah, M. J. (1988) Is visual imagery really visual? Overlooked evidence from neuropsychology. *Psychological Review* 95, 307–17.

p. 65 . . . *that person's voice*. McGuire, P. K. *et al.* Functional anatomy of inner speech and auditory verbal imagery. *Psychological Medicine* 26, 29–38.

p. 67 . . . *without touching them*. Kerr, N. H. The role of vision in "visual imagery" experiments: Evidence from the congenitally blind. *Journal of Experimental Psychology*: General 112, 265–77.

p. 69 . . . *as sighted people*. Op cit.

p. 69 . . . *she could see*. Goldenberg, G., *et al.* (1995) Imagery without perception: A case study of anosognosia for cortical blindness. *Neuropsychologia* 33, 1373–82.

p. 70 . . . *they went blind*. Buchel, C., *et al.* (1998) Different activation patterns in the visual cortex of late and congenitally blind subjects. *Brain* 409–19.

p. 72 . . . *particular imagery method*. Emmorey, K., *et al.* (1998) Mental rotation within linguistic and non-linguistic domains in users of American Sign Language. *Cognition* 68, 221–46.

p. 72 . . . *for spatial patterns*. Emmorey, K. (1998) The impact of sign language use on visuospatial cognition. In M. Marschark and M. D. Clark, *Psychological Perspectives on Deafness*, Vol. 2, New Jersey: Lawrence Erlbaum Associates, 19–52.

p. 73 . . . *less experienced colleagues*. McGuire, E., *et al.* (2000) Navigation-related structural change in the hippocampi of taxi drivers. *Proceedings of the National Academy of Sciences* 97: (8) 4398–403, 11 April.

p. 75 . . . *practice these separately*. Wallace, B., and Hofelich, B. G. (1992) Process generalization and the prediction of performance on mental imagery tasks. *Memory and Cognition* 20, 695–704.

Chapter 5

p. 77 . . . *sweeps of stone*. Wilson, B. A., Baddeley, A. D., and Young, A. W. (1999) LE, a person who lost her 'mind's eye.' *Neurocase* 5, 119–27.

p. 80 . . . *frontotemporal dementia*. Miller, B. L., *et al.* (2000) Functional correlates of musical and visual ability in frontotemporal dementia. *British Journal of Psychiatry* 176, 458–63.

p. 84 . . . *"my elements of thought are . . . images . . ."* Einstein, A. (1952) Letter to Jacques Hadamard. In B. Ghiselin (ed.) *The Creative Process*, 43–4. Berkeley: University of California Press.

p. 85 . . . *fundamental thought process*. Finke, R. A., Pinker, S., and Farah, M. J. (1989) Reinterpreting visual patterns in mental imagery. *Cognitive Science* 13, 51–78.

p. 85–86 . . . *new abstract inventions*. Finke, R. A., and Slayton, K. (1988) Explorations of creative visual synthesis in mental imagery. *Memory and Cognition* 16, 252–7.

p. 87 . . . *over several years*. Getzels, J. W., and Csikszentmihalyi, M. (1976) *The Creative Vision: A Longitudinal Study of Problem-Finding in Art*. New York: John Wiley and Sons.

p. 89 . . . *quite simple methods*. Finke, R. A., Ward, T. B., and Smith, S. M. (1996) *Creative Cognition: Theory, Research and Application*. Cambridge: MIT Press.

p. 90 . . . *through the problems*. Schooler, J. W., Ohlson, S., and Brooks, K. (1993) Thoughts beyond words: When language overshadows insight. *Journal of Experimental Psychology* (General), 122, 166–83.

p. 90 . . . *type of problem*. Kaufmann, G. (1979) *Visual Imagery and Its Relation to Problem Solving*. Bergen, Germany: Universitetsforlaget.

p. 91 . . . *type of thought process*. Schooler, J. W., and Melcher, J. (1995) The ineffability of insight. In S. M. Smith, T. B. Ward, and R. A. Finke (1995) *The Creative Cognition Approach*. Cambridge: MIT Press, 97–133.

p. 92 . . . *can come up with*. Finke, R. A., Ward, T. B., and Smith, S. M. (1996) *Creative Cognition*. Cambridge: MIT Press, 153.

p. 93 . . . *Remote Associates Test*. Mednick, S. A. (1962) The associative basis of the creative process. *Psychological Review* 69, 220–32.

p. 93 . . . *the last three*. Smith, S. M., and Blankenship, S. E. (1991) Incubation and the persistence of fixation in problem solving. *American Journal of Psychology* 104, 61–87.

p. 94 . . . *back of the book. water, pick, skate*: answer *ice; mouse, blue, cottage*: answer *cheese*; and *river, note, blood*: answer *bank*.

p. 96 . . . *by IQ tests*. Sternberg, R. J. (1999) *Handbook of Creativity*. Cambridge: Cambridge University Press.

p. 96 . . . *level of eminence*. Simonton, D. K. (1976) Biographical determinants of achieved eminence. *Journal of Personality and Social Psychology* 33, 218–26.

p. 96–97 . . . *when they leave school*. Nickerson, R. S. (1999) Enhancing Creativity. In R. J. Sternberg (ed.) *Handbook of Creativity*. Cambridge: Cambridge University Press, 392–430.

p. 98 . . . *longer term, it seems*. Baer, J. M. (1988) Long-term effects of creativity

training with middle-school students. *Journal of Early Adolescence* 8, 183–93.
p. 98 . . . *London, for instance.* Adey, P., and Shayer, M. (1993) An exploration of long-term transfer effects following an extended intervention program in the high school science curriculum. *Cognition and Instruction* 11, 1–29.

Chapter 6

p. 104 . . . *Rotman Research Institute in Toronto.* Levine, B., *et al.* (1998) Episodic memory and the self in a case of isolated retrograde amnesia. *Brain* 121, 1951–73.

p. 107 . . . *almost always involved.* Brewer, W. F. (1988) A qualitative analysis of the recalls of randomly sampled autobiographical events. In M. M. Gruneberg, P. E. Morris, and R. N. Sykes (eds.) *Practical Aspects of Memory.* Vol. 1. Chichester, England: John Wiley and Sons, 263–8.

p. 107 . . . *for the first.* Williams, J. M. G., *et al.* (1999) The effects of imageability and predictability of cues in autobiographical memory. *Quarterly Journal of Experimental Psychology* 52A, 555–79.

p. 110 . . . *an internal request.* Tomita, H., *et al.* (1999) Top-down signals from prefrontal cortex in executive control of memory retrieval. *Nature* 401, 699–703.

p. 111–112 . . . *tell wines apart.* Melcher, J. M., *et al.* (1996) The misremembrance of wines past: Verbal and perceptual expertise differentially mediate verbal overshadowing of taste memory. *Journal of Memory and Language* 35, 231–45.

p. 113 . . . *as a word.* Paivio, A., and Csapo, K. (1973) Picture superiority in free recall. *Cognitive Psychology* 5, 176–206.

p. 114 . . . *the word* dog. The various studies described are reviewed in R. L. Buckner *et al.* (1999) Frontal cortex contributes to human memory formation. *Nature Neuroscience* 2, 311–4.

p. 115 . . . *remember as well.* Grady, C. L., *et al.* (1995) *Science* 269: (5221) 218–21, 14 July.

p. 116 . . . *as they memorize.* Buckner, R. Paper presented at the International Congress on Psychology, Stockholm, 2000.

p. 118 . . . *Alan Paivio.* Paivio, A. (1971) *Imagery and Verbal Processes.* New York: Holt, Rinehart and Winston.

p. 119 . . . *are easily visualizable.* Riding, R. J., and Taylor, E. M. (1976) Imagery performance and prose comprehension in seven-year-old children. *Education Studies* 2, 21–7.

p. 121 . . . *the same speed.* Denis, M. (1982) Imaging while reading text: A study of individual differences. *Memory and Cognition* 10, 540–5.

p. 122 . . . *visuospatial memory tests.* Vicari, S., *et al.* (1996) Memory abilities in children with Williams syndrome. *Cortex* 32, 503–14.

p. 123 . . . *in their childhood.* Garry, M., *et al.* Imagination inflation: Imagining a childhood event inflates confidence that it occurred. *Psychonomic Bulletin and Review*, in press.

p. 124 . . . *their children's lives.* Hyman, I.E., and Pentland, J. (1996) The role of mental imagery in the creation of false childhood memories. *Journal of Memory and Language* 35, 101–17.

p. 135 . . . *you are interested.* Every country has its own set of memory gurus and memory books. One example in the UK, for instance, is a series of books by Tony Buzan—for example, *Master Your Memory* (1989) BBC Books.

Chapter 7

p. 141 . . . *at the beginning.* Marzillier, J. S., *et al.* (1979) Self-report and physiological changes accompanying repeated imagining of a phobic scene. *Behaviour Research and Therapy* 17, 71–7.

p. 141 . . . *of the brain.* Fredrikson, M., *et al.* (1997) Cerebral blood flow during anxiety provocation. *Journal of Clinical Psychiatry* 58, 16–21.

p. 142 . . . *50 percent.* Baum, A., *et al.* (1993) Control and intrusive memories as possible determinants of chronic stress. *Psychosomatic Medicine* 55, 274–86.

p. 145 . . . *ways just described.* Pitman, R. K. (1987) Psychophysiologic assessment of post-traumatic stress disorder imagery in Vietnam combat veterans. *Archives of General Psychiatry* 44, 970–5.

p. 146 . . . *much less traumatized.* Stutman, R. K., and Bliss, E. L. (1985) Post-traumatic stress disorder, hypnotizability and imagery. *American Journal of Psychiatry* 142, 741–3.

p. 147 . . . *to a minimum.* Borcovec, T. D., and Hu, S. (1990) The effect of worry on cardiovascular response to phobic imagery. *Behaviour Research and Therapy* 28, 69–73.

p. 148–149 . . . *picture the scene.* Op cit.

p. 149–150 . . . *guilt-driven activity.* Shin, L. M., *et al.* (2000) Activation of anterior paralimbic structures during guilt-related script-driven imagery. *Biological Psychiatry* 48, 43–50.

p. 151 . . . *the frontal lobes.* Baker, S. C., *et al.* (1997) The interaction between mood and cognitive function studied with PET. *Psychological Medicine* 27, 565–78.

p. 151 . . . *left frontal lobe*. Bartolic, E. I., *et al*. (1999) Effects of experimentally induced emotional states on frontal lobe cognitive task performance. *Neuropsychologia* 37, 677–83.

p. 151 . . . *its bloody body*." Vrana, S. R. (1993) The psychophysiology of disgust: Differentiating negative emotional contexts with facial EMG. *Psychophysiology* 30, 279–86.

p. 152 . . . *nonsexual activities!* Rauch, S. L., *et al*. (1999) Neural activation during sexual and competitive arousal in healthy men. *Psychiatry Research—Neurimaging* 91, 1–10.

p. 153 . . . *in his students*. Rachman, S. (1966) Sexual fetishism: An experimental analogue. *Psychological Records* 16, 293–6.

p. 154 . . . *can kill them*. Hinson, R. E., and Siegel, S. (1980) The contribution of Pavlovian conditioning to ethanol tolerance and dependence. In H. Rigter and J. Crabbe, *Alcohol Tolerance and Withdrawal*, New York: Elsevier.

p. 159 . . . *with the drug*. Sinha, R., *et al*. (1999) Stress-induced craving and stress response in cocaine-dependent individuals. *Psychopharmacology* 142, 343–51.

p. 162 . . . *in the laboratory*. Jones, T., and Davey, G. (1990) The effects of cued UCS rehearsal on the retention of differential 'fear' conditioning. *Behaviour Research and Therapy* 28, 159–64.

p. 165 . . . *your mind's eye*. . . . Challis, G. B., and Stam, H. J. (1992) A longitudinal study of the development of ANV in cancer chemotherapy patients: The role of absorption and autonomic arousal. *Health Psychology* 11, 181–9.

p. 166 . . . *in real life*. Wade, T. C., *et al*. (1977) Imaginal correlates of self-reported fear and avoidance behavior. *Behaviour Research and Therapy* 15, 17–22.

p. 168 . . . *obsessive-compulsive disorder*. Foa, E. B., *et al*. (1980) Effects of imaginal exposure to feared disasters in obsessive-compulsive checkers. *Behaviour Research and Therapy* 18, 449–55.

p. 172 . . . *endure real shocks*. Yaremko, R. M., and Butler, M. C. (1975) Imaginal experience and the attenuation of the galvanic skin response to shock. *Bulletin of the Psychonomic Society* 5, 317–8.

p. 172 . . . *the outside world*. Marzillier, J. S., *et al*. (1979) Self-report and physiological changes accompanying repeated imagining of a phobic scene. *Behaviour Research and Therapy* 17, 71–7.

Chapter 8

p. 175 . . . *stopped taking the drugs*. This case is an imaginative reconstruction of a number of real cases reported in a study of the effects of imagery and hypnosis on herpes simplex patients in a London hospital. Many thanks to Professor John Gruzelier of Imperial College for giving me additional details of the imagery methods used in the study. The study was published as: Fox, P. A., *et al.* (1999) Immunological markers of frequently recurrent genital herpes simplex virus and their response to hypnotherapy. *International Journal of STD and AIDS* 10, 730–4.

p. 176–177 . . . *the molecules themselves*. Forlenza, M. J., and Baum, A. (2000) *Current Opinion in Psychiatry* 13, 639–45.

p. 177 . . . *for an MI*. Lewin, B., *et al.* (1992) *Lancet* 339, 1036–40.

p. 177 . . . *still being given*. Ader, R., *et al.* (1992) *International Journal of Immunopharmacology* 14: (3) 323–7, April.

p. 178 . . . *did change immunity*. Buskekirschbaum, A., *et al.* (1994) Conditioned manipulation of natural killer (NK) cells in humans using a discriminative learning protocol. *Biological Psychology* 38, 143–55.

p. 178–179 . . . *to the tuberculin*. Smith, G. R., and McDaniels, S. M. (1983) Psychologically mediated effect on the delayed hypersensitivity reaction to tuberculin. *Psychosomatic Medicine* 45, 65–77.

p. 180 . . . *the drug itself*. Ader, R., and Cohen, N. (1982) Behaviorally conditioned immunosuppression and murine systemic lupus erythematosus. *Science* 215, 1534–7.

p. 180 . . . *can be conditioned*. Irie, M., *et al.* (2000) Classical conditioning of oxidative DNA damage in rats. *Neuroscience Letters* 288, 13–16.

p. 182 . . . *chemotherapy is stopped*. Stockhorst, U., *et al.* (2000) Anticipatory symptoms and anticipatory immune responses in pediatric cancer patients receiving chemotherapy: Features of a classically conditioned response? *Advanced Drug Delivery Reviews* 42, 198–218.

p. 182 . . . *this conditioned response*. Challis, G. B., and Henderikus, J. S. (1992) A longitudinal study of the development of anticipatory nausea and vomiting in cancer chemotherapy patients: The role of absorption and autonomic perception. *Health Psychology* 11, 181–9.

p. 182 . . . *types of hypnosis*. Redd, W. H., *et al.* (1982) Hypnotic control of anticipator emesis in patients receiving cancer chemotherapy. *Journal of Consulting and Clinical Psychology* 50, 14–19.

p. 183 . . . *hot and spicy substances*. Lutgendorf, S., *et al.* (2000) Effects of relaxation and stress on the capsaicin-induced local inflammatory response. *Psychosomatic Medicine* 62, 524–34.

p. 185 . . . *of psoriasis sufferers*. Zacharie, R., *et al.* (1996) Effects of psychologic intervention on psoriasis: A preliminary report. *Journal of the American Academy of Dermatology* 34, 1008–15.

p. 187 . . . *by imagery exercises*. Olness, K., *et al.* (1998) Mast cell activation in children with migraine before and after training in self-regulation. *Headache* 39, 101–7.

p. 190 . . . *the immune system*. Zacharie, R., *et al.* (1990) Effect of psychological intervention in the form of relaxation and guided imagery on cellular immune function in normal healthy subjects. *Psychotherapy and Psychosomatics* 54, 32–9.

p. 191 . . . *and control groups*. Fawzy, F. I., *et al.* (1993) Malignant-melanoma: Effects of an early structured psychiatric intervention, coping and affective state on recurrence and survival 6 years later. *Archives of General Psychiatry* 50, 681–9.

p. 191 . . . *advanced breast cancer*. Spiegel, D., *et al.* (1989) Effect of psychosocial treatment on survival of patient with metastatic breast cancer. *Lancet* 2, 888–90.

Chapter 9

p. 196 . . . *figure skater Nancy Kerrigan*. See Moran, A. P. (1996) *The Psychology of Concentration in Sports Performers*. Hove, England: Psychology Press.

p. 196 . . . *the mind's eye*. See *Newsweek*, 25 Sept 2000, 75–6, for these interviews.

p. 196 . . . *are intimately linked*. Deschaumes, C., *et al.* (1991) Relationship between mental imagery and sporting performance. *Behavioural Brain Research* 45, 29–36.

p. 197–198 . . . *during mental practice*. Roure, R., *et al.* (1999) Imagery quality estimated by autonomic response is correlated to sporting performance enhancement. *Physiology and Behavior* 66, 63–72.

p. 198 . . . *imagine the exercise*. Kim, J., *et al.* (1998) Visual, auditory and kinesthetic imagery on motor learning. *Journal of Human Movement Studies* 34, 159–74.

p. 201 . . . *of training sessions*. Yue, G., and Cole, K. J. (1992) Strength increases from the motor program: Comparison of training with maximal voluntary and imagined muscle contractions. *Journal of Neurophysiology* 67, 1114–23.

p. 202 . . . *simple computer joystick*. Stephan, K. M., *et al.* (1995) Functional anatomy of the mental representation of upper extremity movements in healthy subjects. *Journal of Neurophysiology* 73, 373–86.

p. 203 . . . *changing the brain*. Pascual-Leone, A., *et al.* (1995) Modulation of muscle responses evoked by transcranial magnetic stimulation during the acquisition of new fine motor skills. *Journal of Neurophysiology* 74, 1037–45.

p. 204–205 . . . *the other hand.* For a good review of several of these mental imagery studies, see M. Jeannerod and J. Decety (1995) *Current Opinion in Neurobiology* 5, 727–32.

p. 205 . . . *in his mind.* The source of the information about the training of doctors and surgeons, as well as the anecdote about Glenn Gould, is J. G. Des Coteaux and H. Leclere. (1995) Learning surgical technical skills. *Canadian Journal of Surgery* 38, 33–8.

p. 206 . . . *blow to the head.* Miltner, R., *et al.* (1999) Bewegungsvorstellung in der Therapie von Patienten mit Hirninfarkt. *Neurological Rehabilitation* 5, 66–72.

Chapter 10

p. 210 . . . *inside their own heads.* Szechtman, H., *et al.* (1998) Where the imaginal appears real: A positron emission tomography study of auditory hallucinations. *Proceedings of the National Academy of Science*, 95, 1956–60.

p. 211 . . . *likely to be.* Lynn, S. J., and Rhue, J. W. (1986) The fantasy-prone person: Hypnosis, imagination and creativity. *Journal of Personality and Social Psychology* 51, 404–8.

p. 212 . . . *more creative than average.* Op cit.

p. 212 . . . *color in the gray.* Kosslyn, S., *et al.* (2000) Hypnotic visual illusion alters color processing in the brain. *American Journal of Psychiatry* 157, 1279–84.

p. 213 . . . *his left leg.* Halligan, P. W., *et al.* (2000) Imaging hypnotic paralysis: Implications for conversion hysteria. *Lancet* 355, 986–7.

p. 213 . . . *researchers in Montreal.* Rainville, P., *et al.* (1999) Cerebral mechanisms of hypnotic induction and suggestion. *Journal of Cognitive Neuroscience* 11, 110–25.

p. 215 . . . *had been painless.* See Hilgard, E. R., and Hilgard, J. R. (1983) *Hypnosis and the Relief of Pain.* Los Altos: William Kaufmann, p. 63.

p. 216 . . . *Hypnosis and the Relief of Pain. . . .* Op cit.

p. 217 . . . *the mind's eye.* Lambert, S. A. (1996) The effects of hypnosis-guided imagery on the post-operative course of children. *Journal of Developmental and Behavioral Pediatrics* 17, 307–10.

p. 217 . . . *cancer hospital in Seattle.* Syrjala, K. L. (1995) Relaxation and imagery and cognitive-behavioral training reduce pain during cancer-treatment: A controlled clinical trial. *Pain* 63, 189–98.

p. 218 . . . *"recovered" under hypnosis*. Persinger, M. A. (1992) Neuropsychological profiles of adults who report sudden remembering of early-childhood memories. *Perceptual and Motor Skills* 75, 259–66.

Chapter 11

p. 221 . . . *the parietal lobes*. Stickgold, R. (1998) Sleep: Off-line memory processing. *Trends in Cognitive Science* 12, 484–92.

p. 222 . . . *the dream run*. Hobson, J. A., Pace-Schott, E., and Stickgold, R. Dreaming and the brain: Towards a cognitive neuroscience of conscious states. *Behavioral and Brain Sciences*. In press.

p. 223 . . . *for a touchdown*. Boeve, B. F., *et al*. (1998) REM sleep behavior disorder and degenerative dementia: An association likely reflecting Lewy body disease. *Neurology* 51, 363–70.

p. 224 . . . *beat the nightmares too*. Krakow, B., *et al*. (1995) Imagery rehearsal treatment for chronic nightmares. *Behaviour Research and Therapy* 33, 837–43.

p. 224 . . . *that traumatic event*. Nader, K. (1996) Children's traumatic dreams. In D. Barrett (ed.) *Trauma and Dreams*. Cambridge: Harvard University Press.

p. 225 . . . *in the future*. Revonsuo, A. (2000) The reinterpretation of dreams: An evolutionary hypothesis of the function of dreaming. *Behavioral and Brain Sciences* online.

p. 225 . . . *beatings, and killings*. Bilu, Y. (1989) The other as nightmare: The Israeli–Arab encounter as reflected in children's dreams in Israel and the West Bank. *Political Psychology* 10, 365–89.

p. 226 . . . *overnight—effortless—learning*. Karni, A., *et al*. (1994) Dependence on REM sleep of overnight learning of a perceptual skill. *Science* 265, 679–82.

p. 226–227 . . . *in the dream*. Louie, K., and Wilson, M. A. (2001) Temporally structured replay of awake hippocampal ensemble activity during rapid eye movement sleep. *Neuron* 29:(1) 145–56, January.

p. 228 . . . *when you are dreaming*. Hobson, J. A., Pace-Schott, E., and Stickgold, R. (2000) Dreaming and the brain: Towards a cognitive neuroscience of conscious states. *Behavioral and Brain Sciences* online.

p. 228 . . . *less frightening events*. Zadra, A. L., and Pihl, R. O. (1997) Lucid dreaming as a treatment for recurrent nightmares. *Psychotherapy and Psychosomatics* 66, 50–5.

p. 229 . . . *apparently fast asleep*. Mason, L. I., *et al*. (1997) Electrophysiological correlates of higher states of consciousness during sleep in long-term practitioners of the transcendental meditation program. *Sleep* 20, 102–10.

Chapter 12

p. 233 . . . *among men, that is*. Neeleman, J., and Lewis, G. (1990) Suicide, religion, and socioeconomic conditions: An ecological study in 26 countries. *Journal of Epidemiology and Community Health* 53:(4) 204–10.

p. 234 . . . *brain had decreased*. Newberg, A., *et al*. (1995) HMPAO-SPECT Imaging during intense Tibetan Buddhist meditation. *Biological Psychiatry* 37, 619.

p. 235 . . . *as they meditated*. Lou, H. C., *et al*. (1999) A 15-O-H20 PET study of meditation and the resting state of normal consciousness. *Human Brain Mapping* 7, 98–105.

p. 238 . . . *this was the afterlife*. Young, A. W., Robertson, I. H., *et al*. Cotard delusion after brain injury. *Psychological Medicine* 22, 799–804.

p. 239–240 . . . *right hemisphere semantic systems*. Brugger, P., *et al*. (1993) 'Meaningful' patterns in visual noise: Effects of lateral stimulation and the observer's belief in ESP. *Psychopathology*, 26, 261–5.

p. 243 . . . *"mindfulness-based cognitive therapy."* Teasdale, J. D., *et al*. (2000) Prevention of relapse/recurrence in major depression by mindfulness-based cognitive therapy. *Journal of Consulting and Clinical Psychology* 68: (4) 615–23.

p. 249 . . . *realm of learning*. Gould, S. J. (2001) *Rocks of Ages*. New York: Ballantine Books.

Index